CONTENTS

Introduction

What a Lot of Rubbish is the fourteenth volume in the series: **Issues For The Nineties**. The aim of this series is to offer up-to-date information about important issues in our world.

What a Lot of Rubbish looks at recycling, energy efficiency and energy from waste.

The information comes from a wide variety of sources and includes:
Government reports and statistics
Newspaper reports and features
Magazine articles and surveys
Literature from lobby groups
and charitable organisations.

It is hoped that, as you read about the many aspects of the issues explored in this book, you will critically evaluate the information presented. It is important that you decide whether you are being presented with facts or opinions. Does the writer give a biased or an unbiased report? If an opinion is being expressed, do you agree with the writer?

What a Lot of Rubbish offers a useful starting-point for those who need convenient access to information about the many issues involved. However, it is only a starting-point. At the back of the book is a list of organisations which you may want to contact for further information.

*****

Editor

Craig Donnellan

Independence
Educational Publishers
Cambridge

First published by Independence
PO Box 295
Cambridge CB1 3XP

© Craig Donnellan 1997

British Library Cataloguing in Publication Data
What a Lot of Rubbish– (Issues for the Nineties Series)
I. Donnellan, Craig II. Series
363.7'28

ISBN 1 86168 022 8

Printed in Great Britain
City Print Ltd
Milton Keynes

Typeset by
Claire Boyd

Cover
The illustration on the front cover is by
Andrew Smith / Folio Collective.

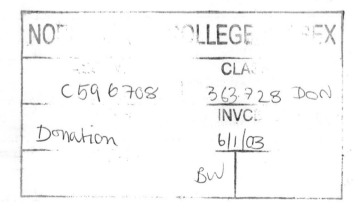

What did you throw out this week?

A throw-away society

A modern consumer society like ours can become a throw-away society.

Every year we generate an incredible 435 million tonnes of waste in the United Kingdom. Most of this has to be disposed of through landfill and incineration.

We cannot stop the production of waste – but we can reduce its impact on the environment.

Only by taking more responsibility for our waste – as individuals and as communities – can we ensure that our environment is protected both now and for future generations.

The key is to reduce the quantity of waste we produce, and to recycle or compost not less than 25 per cent of household waste by the end of the century. The current average is only 6 per cent, so we have a long way to go.

Everyone has an important contribution to make towards achieving these aims.

Start making your contribution now. See what you can do straight-away as a householder and as a consumer.

Every nine months we produce enough waste in the United Kingdom to fill Lake Windermere.

What you can do as a householder

The four 'Rs'

Reduce the amount of your waste that needs to be disposed of – by composting, if you can.

Re-use things – such as bottles, carrier bags and refillable containers.

Recycle things – use local recycling facilities for such items as bottles, cans, paper, plastic and rags.

Repair things – don't always replace damaged goods with new ones.

What you can do as a consumer

- Buy re-usable products and products with either no or very little packaging.
- When you are buying think about how long products will last, their efficiency, and the impact of throwing them away.
- Buy products marked with the European Community eco-label – for example, washing machines and dishwashers.
- Avoid, where possible, buying products containing hazardous materials, such as lead acid batteries and certain bleaches. There may be a non-hazardous alternative.
- Buy products that are energy efficient – look for official energy labels or other energy efficiency information on appliances.

Remember

You can make a difference

- by reducing your waste
- by re-using or recycling your waste
- by starting a compost heap
- by buying carefully.

It's time to start making your waste work.

WE'VE BEEN COMING HERE FOR OUR PICNICS FOR YEARS!

Going for Green

Making a world of difference – together

You can find out more about how to make changes in your lifestyle that will improve and preserve the environment by contacting the Government-backed national campaign Going for Green. Write to: Going for Green, PO Box 2100, Manchester M60 3GN.

Whatever you threw out this week, it wasn't all waste.

Think about what you did throw out. It is likely that you are wasting things that could be recycled or re-used.

Think about the amount of waste you produce. It could be reduced.

For the sake of the environment, we need to reduce the amount of waste we produce and to make better use of the waste we do produce.

Some of the things that you can re-use, recycle or compost:

Newspapers and magazines
Glass bottles
Plastics
Tin cans
Aluminium cans
Garden and kitchen waste
Start watching your waste now.

Information

Your council's recycling officer will be able to give you advice about the recycling facilities in your area. And many voluntary, community and environmental groups run schemes to help you make the most of your waste – your local council or library may be able to help you to find out about them.

Making Waste Work – A strategy for sustainable waste management in England and Wales, a 16-page summary of *Making Waste Work*, the Government's White Paper on sustainable waste management, is available free from the Department of the Environment, Publications Despatch Centre, Blackhorse Road, London SE99 6TT (Fax: 0181 694 0099). Please quote reference number 95 EP 130.

The White Paper *Making Waste Work* is available from HMSO at £16. Mail orders from: HMSO Publications Centre, PO Box 276, London SW8 5DT. Telephone orders: 0171 873 9090. General enquiries: 0171 873 0011. Please quote ISBN No: 0-10-130402-1.

Produced by the Department of the Environment
© Crown copyright 1997
February, 1997

Research and surveys

Information from Tidy Britain Group (TBG)

Research and surveys form the backbone of much of the work carried out by Tidy Britain Group. Identifying problems locally and nationally with regard not only to litter, but also to its associated problems – fly-posting, graffiti and dog fouling – allows the Group to formulate strategies to deal with these issues.

Uniquely, the Group has also monitored cleanliness standards in over 80 major urban areas throughout the UK. The Group has become the authoritative voice on cleanliness standards. Its standards are recognised nationally by the Audit Commission who are recommending that TBG's methods are adopted by local authorities in monitoring their own cleanliness standards.

In 1996, TBG published its first-ever State of the Nation report, which combined the results of several major pieces of research conducted over the previous 18 months.

State of the nation

This report highlighted the results of research which had examined cleanliness standards, including the incidence of dog fouling, graffiti and fly-posting, in over 80 city and town centres throughout the UK. It also included the results of a survey into public attitudes towards litter and the local environment.

The report clarified major issues in relation to local environmental problems and helped to identify areas which need further research and development.

State of cleanliness in major urban areas in the UK

- 10% of sites were litter-free (Grade A)
- 75% of sites had scattered small items of litter (Grade B)
- 12% of sites had small accumulations of litter (Grade C)
- 3% of sites had large accumulations of litter (Grade D)

Smoking-related litter

It also became clear from this survey that the most commonly found item of litter was smoking-related litter, which affected 86% of sites overall. It was often the sole reason for down-grading sites from A to B and therefore significantly affected the national cleanliness standards.

Based on this information, TBG is currently conducting a series of surveys into smoking-related litter – the effects of no-smoking policies in the workplace, the disposal habits of people who smoke and the distribution and design of bins for disposal. The results of this research should be published in Autumn 1996.

Local environmental quality

Problems like dog fouling, graffiti and fly-posting tend to be associated with litter – their presence is examined in all TBG cleanliness surveys. The State of the Nation report, for the first time, gave an accurate picture of how these environmental problems affected the UK's major urban areas.

- Dog fouling affected 7.25% of sites nationwide.
- Graffiti affected 15.88% of sites nationwide.
- Fly-posting affected 10.42% of sites nationwide.

The information compiled in such reports assists local authorities in making decisions as to the best use of resources. Tackling issues which affect the quality of the local environment helps to provide a better quality of life for residents and visitors alike.

Public attitudes towards local environmental problems

Dog fouling was named by 80% of people questioned as the local environmental problem which caused them most concern. Although affecting only 7.25% of sites nationwide it can, in some urban areas, affect up to 40% of surveyed sites.

This work, in association with earlier TBG research, was used to support lobbying for new legislation, which would help local authorities to deal with the problem of dog fouling. This has resulted in the Dogs (Fouling of Land) Act which received Royal Assent earlier in the year and will become law in Autumn 1996.

TBG has also produced an action/information pack, *Picking Up the Problem*, based on the expertise and best practice of local authorities,

and deals with other aspects of dog hygiene, such as campaigning and education. This pack is designed specifically for local authorities, and will help to complement the new legislative powers.

People action

Of the people questioned, 30% had taken action themselves against litter. Of these, the majority (63%) had cleaned up litter themselves. However, 19% had approached their local council, while 14% had tackled litterbugs directly.

93% of people were aware that it was an offence to drop litter in a public place. However, only 3% thought the maximum fine for littering was over £1,000 – the maximum fine is actually £2,500.

Top litter items

Smoking-related litter affected over 80% of the surveyed sites, followed by sweet wrappers at 40%, drinks cans at 26% and fast-food packaging at 12%. However, it was the latter – fast-food packaging – which the public considered to be the most offensive type of litter – over 70% of people designated it such.

Future research

Keeping in touch with problem areas identified in its own research, and conscious of potential waste management issues raised by changes in legislation etc, TBG research in the coming year will include smoking-related litter, fly-tipping, cleanliness standards in rural areas, clinical waste disposal and a repeat of the UK City Survey.

• The above is an extract from the Annual Report 1995/96 of Tidy Britain Group. See page 39 for address details.   © *Tidy Britain Group*

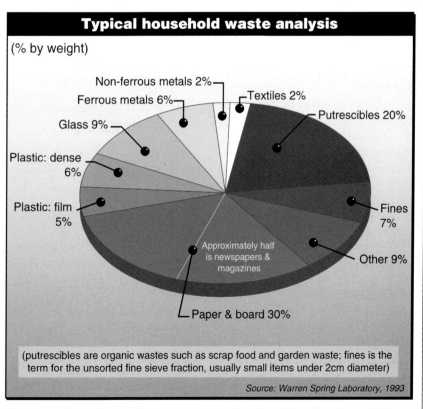

Typical household waste analysis

(% by weight)

Non-ferrous metals 2%
Ferrous metals 6%
Glass 9%
Plastic: dense 6%
Plastic: film 5%
Textiles 2%
Putrescibles 20%
Fines 7%
Other 9%
Approximately half is newspapers & magazines
Paper & board 30%

(putrescibles are organic wastes such as scrap food and garden waste; fines is the term for the unsorted fine sieve fraction, usually small items under 2cm diameter)

Source: Warren Spring Laboratory, 1993

Don't throw it all away

Friends of the Earth's guide to waste reduction and recycling

The environmental impacts of waste

Our 'throwaway' culture means that we use up large quantities of raw materials and energy. Burying and burning all our waste creates serious pollution problems.

Wasted resources

Throwing so much away is a senseless misuse of the Earth's precious resources! One way to save resources is to use waste itself as a raw material. Even better is to waste less in the first place.

Some raw materials are 'finite' – which means that reserves will simply run out if we continue to squander them to make throwaway goods. Recycling products made from finite resources like plastics (made from oil) and aluminium (made from bauxite) can help to conserve supplies. Reducing unnecessary packaging and making things to last also help.

Reducing and recycling waste also relieves the environmental damage done to wildlife habitats when raw materials are mined, quarried or 'harvested'.

- Every tonne of recycled glass saves 1.2 tonnes of sand, soda ash and limestone.
- Recycling one tonne of steel saves 900 kilograms of raw materials.
- It takes 10 to 18 trees to make one tonne of paper – so using less paper and recycling waste reduces the need for intensive forestry plantations and for felling trees in ancient natural forests.

Wasted energy

Using scrap materials to make new products takes less energy too. Saving energy helps to reduce all the problems associated with energy generation: acid rain, smog, radioactive pollution from nuclear accidents and waste, the flooding of valleys for huge hydro-electric power schemes and the threat of climate change.

Potential energy savings from using scrap materials instead of virgin ones:

Glass	22%
Paper	70%
Aluminium	96%
Steel	74%
Polyethylene	97%

Pollution

Disposing of all the waste we create is a problem in itself. Both burying and burning waste can cause hazards to human health and to the environment. It is not only 'toxic waste' that is a hazard: household waste can also produce polluting liquids and explosive gases when buried in the ground and harmful fumes when incinerated.

Thousands of landfill sites around the country are generating polluting gas and liquids and threatening to contaminate drinking water supplies.

We are running out of sites where waste can be landfilled, especially close to cities where most of the waste is produced. So waste has to be transported further and further away from its source, adding to the environmental damage caused by its disposal.

Reduce, re-use, recycle

The best way of reducing the problems caused by waste is to produce less waste. Where waste cannot be avoided, it should be re-used or recycled wherever possible.

Waste reduction: throw away the throwaway society

We are producing more and more waste. Between 1985 and 1989 the amount of household waste produced per person in the UK rose by 5 per cent, whereas in some other countries, including Canada and Norway, the amount of waste each person produced actually fell.

Friends of the Earth has asked the Government to set targets for reducing waste and to develop strategies for meeting them.

We could all create less rubbish by refusing to buy goods with unnecessary packaging and by buying

clothes and household goods that are designed to last several years rather than a few months. Industry too can adapt its manufacturing processes so that they are less wasteful.

Re-use: bring back the bring-back

Re-using something takes less energy and creates less pollution than reprocessing it to make a new product.

Many items – like clothes, furniture and household goods – get thrown away simply because they are old or broken. They could be repaired, refurbished and re-used.

Products like food, drinks, household cleaning products and toiletries, could be sold in re-usable containers instead of one-trip packaging. Studies have shown that re-usable packaging systems use fewer materials and less energy than one-trip packs and create less pollution and waste. Friends of the Earth is calling on the big supermarket chains to introduce re-usable packaging schemes.

Recycling: once is not enough

If waste cannot be re-used it should be recycled. Over half of the rubbish in our dustbins could be recycled, but only between 2.5 and 4.5 per cent of it is.

Glass, plastic and metals can be melted and reshaped. Paper can be pulped and made into new paper. Organic waste can be composted. Textiles can be unwoven and respun into new cloth. But the enormous potential is not being met.

In 1990 the Government set a national target – to recycle 25 per cent of household waste by 2000. This is a massive increase on the amount we manage at the moment!

Local authorities will play an important part in making sure we reach this target. The Environmental Protection Act 1990 requires all local authorities which collect waste to write a Recycling Plan, setting out how they intend to recycle waste from homes and shops in their area.

These plans should have been submitted to the Government by 1 August 1992. However, fewer than half of the local authorities who should have prepared plans actually submitted them by the deadline. And

only just over a third told Friends of the Earth that they were planning to meet or exceed the 25 per cent target.

You have a right to see your local council's Recycling Plan. Local authorities will be reviewing their plans in the coming years, and they should take residents' views into account. So let the council know if you think it could do more to increase the amount of waste that is recycled locally.

Recycling collections

Councils around the country are using different methods to collect and sort material for recycling.

Most councils collect material by providing banks or recycling centres where people can bring their recyclables. These are known as 'bring' or 'bank' schemes because people have to bring material to the banks.

Some councils collect recyclable material from people's homes. Recyclables are collected separately from the rest of the refuse – in different dustcarts or in special dustcarts with separate compartments for each material. These are known as 'door-to-door', 'kerbside' or 'collect' schemes, because the material is collected from each household.

Door-to-door schemes are more expensive to set up in the short term. However, comparisons show that a higher proportion of the available material is sent for recycling using door-to-door schemes than where bring schemes are used. In the longer term, door-to-door collections may also lead to financial savings.

• The above is an extract from *Don't throw it all away!*, published by Friends of the Earth. See page 39 for address details.

© Friends of the Earth

The litter collection

Pieces of litter information

Tidy Britain Group surveys show:

- 93% of people questioned were aware that it is an offence to drop litter.
- 80% of people said dog fouling was the local environmental problem causing most concern. Fast-food packaging was regarded as 'very offensive' by 72% of people, with drinks cans, cigarette ends and sweet wrappers close behind.
- Cigarette ends, followed by sweet wrappers and matchsticks, were the most commonly found litter in 1989. When the survey was repeated in 1993, the same items still topped the litter league table.
- 30% of people surveyed thought their area was cleaner than a year ago, whilst 24% thought the UK as a whole was cleaner.
- In a study into attitudes to town and city centres, children aged 9 – 15 perceived litter as the biggest problem. 40% thought that getting rid of litter and graffiti was the best way to improve town centres; more important than better shops and less traffic.
- There are around 6 litter bins per 1,000 population.
- More than 100,000 trolleys go astray from supermarkets each year. Many end up obstructing paths, spoiling beauty spots or polluting rivers and ponds. The cost is some £5 million.
- It can cost a council as much as £14 per kilometre to sweep the streets (depending on method, frequency and condition).
- A typical council spends £1/2 million a year on street cleaning alone (not including cleaning beaches, parks, car parks, markets, special events, etc.).
- Over 20% of local authorities now have litter wardens who can issue £10 fixed penalty fines for littering.
- The average magistrate's court fine for a littering offence is nearly £100.

© Tidy Britain Group, Communications Department
August, 1996

Why recycle?

Information from Waste Watch

Recycling, meaning the processing of manufactured products to provide the material to make new ones, is widely accepted as a good thing to do. What is the basis for this belief?

- There are many environmental advantages in recycling materials. Less energy is used, resulting in less pollution.

- Material recycling reduces the demand for virgin raw materials.

- Since many of these virgin raw materials have to be imported, recycling may also benefit the economy.

- Recycling reduces the amount of waste which needs to be landfilled or incinerated.

- Taking part in recycling enables people to help the environment and gives a sense of individual responsibility for the waste which we all help to create.

However, recycling itself uses energy, may cause pollution and will never wholly replace the need for virgin raw materials. It is better not to create so much waste in the first place. A hierarchy of options for dealing with waste is:

Reduce waste

Re-use waste

Recover materials for recycling

Recover materials for composting

Landfill or incineration with energy recovery

Disposal without energy recovery

(Adapted from *A Waste Strategy for England and Wales*, Department of the Environment/Welsh Office, 1995).

In the Waste Strategy consultation draft it is stated that: 'The Government's overall policy aim for

achieving sustainable waste management is to increase the proportion of waste managed by the options towards the top of the waste hierarchy.'

The key players

In our complex society, effective recycling requires the combines efforts of a number of different people including government, industry and individuals.

Central government

In its White Paper on the Environment (1990), *This Common Inheritance*, the Government set the target of recycling 25% of household waste by the year 2000.

The bulk of United Kingdom waste, about 83%, is still landfilled in some 4,000 controlled sites

---

In 1996 the Government produced *Making Waste Work* a strategy for sustainable waste management in the UK, setting out key policy areas and identifying areas in which the setting of targets could be useful.

The Landfill Tax Regulations and The Finance Act 1996

To encourage alternatives to landfill, with effect from 1 October 1996, waste going to landfill is taxed. Monies collected through these Regulations may be diverted to organisations registered as Environmental Bodies through a credit system.

The Environment Agency

The Environment Agency was set up by the Environment Act 1995 to take on the responsibilities of waste regulation.

Local authorities

The Environment Protection Act (EPA) 1990 introduced a range of measures aimed at stimulating recycling and reducing the amount of waste sent to landfill. All waste collection authorities (local authorities responsible for collecting refuse from households) were required to prepare a recycling plan outlining their strategy for recycling in their area.

The Act also introduced the Recycling Credits Scheme. Recycling Credits are a way of stimulating recycling by passing the savings in collection or disposal costs back to the people organising the recycling. Both Waste Disposal and Waste Collection Authorities have the power to pay Recycling Credits but are not obliged to do so by law.

Many local authorities have plans to exceed the 25% recycling target, and in 1997 several have already done so,

Waste management companies
The waste management industry is now a substantial industry. Increasingly, companies deal with a wide range of waste management activities rather than just one, such as landfill.

Industry and business
A number of industries have a long history of recycling and materials reclamation, for example the scrap metal and textile recycling industries. Newer industries are now developing ways of dealing with waste.

The UK Government encourages industry to agree voluntary methods of meeting EU Directives and recycling targets and several industry groups have been set up such as the Industry Council for Electrical and Electronic Recycling (ICER) and ECTEL, a group currently piloting a mobile phone collection system.

There is increasing interest in examining the whole life of a manufactured product in terms of its environmental impact and this is gradually leading to changes in product design, making items more durable or easier to recycle.

Many companies are now also realising that there are considerable cost and resource savings to be made in improving their own waste management by implementing waste reduction and recycling schemes on their premises.

The European Union
The UK is a member of the European Union and is obliged to follow its directives, including those relating to waste. Recent Directives from the EU include:

Regulation 880/92 on Eco-labelling Schemes (2nd May 1992)
The scheme is a voluntary market mechanism to promote products that have a lesser impact on the environment in certain consumer product groups. The Eco-label is awarded once the environmental impact of a product's whole life cycle (including disposal) has been assessed with respect to other similar products.

EU Directive 94/62 on Packaging and Packaging Waste
The main aims of the Directive are to reduce the overall impact of packaging on the environment; to harmonise Member States existing packaging legislation and to remove trade obstacles which may distort competition. It applies to all packaging, regardless of material and wherever it is used, whether industrial, commercial or shop level. The overall target of the Directive is to ensure the recovery of 50% of packaging across Member States by the year 2001, with a minimum 16% recycling rate for each material.

The Voluntary and community sector
This sector plays a central role in developing locally based action on waste reduction, re-use and recycling programmes.

Individuals
Most householders now participate in recycling schemes run by their local authority or by voluntary groups.

Householders also have part to play in 'closing the loop' and buying products which contain recycled materials as recycling depends on there being markets for the materials collected.

Where waste goes – round, up or down?
There are three main options for dealing with waste – recycling, incineration and landfill. With both incineration and landfill it is possible to recover energy.

Recycling
Potentially recyclable material, such as glass, cans, plastic bottles, textiles or paper, accounts for about 50% by weight of the contents of the average dustbin. Vegetable peelings and other organic material make up a further 20%. Although over 50% of household waste is estimated to be recyclable, at present only 5-6% is recovered. Recycling rates are higher in a number of other European countries; Denmark and Finland, for example, both achieving about 20%.

Incineration
Under 10% of household waste is currently incinerated or burnt. The residue from incineration, which is 25-30% of the original volume and may contain toxic chemicals and heavy metals, still has to be landfilled. Incineration is used more widely in other European countries, many of which have fewer sites suitable for landfill. Over 70% of waste in Denmark, for example, is incinerated, mostly with energy recovery.

Landfill
The bulk of United Kingdom waste, about 83%, is still landfilled in some 4,000 controlled sites. For much of the country landfill has been the cheapest method of waste disposal. This is now changing as higher standards for landfill are now required by the EPA (1990) and costs are rising due to the introduction of the Landfill Tax.

• The above is an extract from *Recycling, Now!*, published by Waste Watch. See page 39 for address details.

Making waste work

An extract from *Making Waste Work: Summary*, published by the Department of the Environment and the Welsh Office

Our targets

In *Making Waste Work*, we have set two primary targets and made a commitment to set a third. To focus attention on the different ways we can meet these targets, we have also set down a range of supporting secondary targets.

Our primary targets are:
- to reduce the proportion of controlled waste going to landfill to 60 per cent by 2005;
- to recover 40 per cent of municipal waste by 2005; and
- by the end of 1998, to set a target for overall waste reduction.

Waste reduction

Waste reduction stands apart from the other options in the waste hierarchy. By avoiding the production of waste, we also avoid the costs associated with the disposal or recovery of waste. Policies for minimising waste are at the heart of *Making Waste Work*.

To this end:
- by the end of 1998, the Government will set a target for overall waste reduction.

We need to be concerned with not only reducing the amount of waste needing to be processed or disposed of, but also reducing the risks represented by waste.

This is why the Government will promote reduction of:
- the quantity of waste produced;
- the hazardousness of waste.

Action can be taken by designers, manufacturers and consumers to reduce both the quantity and hazardousness of waste. Through the work of the Environmental Technology Best Practice Programme and bodies such as the Research Councils, we are learning more and more about the possibilities of waste reduction and the benefits it can bring.

The Environmental Technology Best Practice Programme will be an important means of communicating the waste reduction message to business. The Government has asked Going for Green (the citizens' initiative on sustainable development) to take forward promotion of waste reduction to local authorities and householders.

Why reduce your waste?
You can save:
- on the environmental and economic costs of raw materials;
- on the environmental and economic costs of waste disposal; and
- on the environmental and economic costs of collecting, transporting and processing both raw materials and waste materials.

Re-use

Re-use of many everyday items is taken for granted. The re-use of milk bottles, for example, has long been part of both household and commercial life. However, the growth in consumer choice, changing social patterns and new business practices have led to a decline in re-use in recent years, particularly at the consumer level.

Re-use is not without its costs: in particular, the infrastructure and expense of return, cleansing and refilling, and the costs of redesigning products to make them re-usable. But the benefits will often outweigh the costs and there are signs of renewed interest in re-use.

To promote re-use the Government will:
- encourage businesses and the consumer to do more to realise the economic and environmental gains that can be obtained in this area; and
- continue to work with voluntary groups such as Waste Watch to encourage imaginative and inno-vative local approaches to re-use.

Re-use is a practical option in the home – for example, re-using bottles or buying refillable and repairable products. On a larger scale, a number of waste exchanges currently operate in the UK. Their general aim has been to draw up and circulate lists of types of wastes produced by companies in different locations, so giving other companies the opportunity to use the wastes in place of raw materials.

Why re-use waste?
- You can save on the environmental and economic costs of raw materials.
- You can reduce the amount of waste that needs disposal and save on the associated environmental and economic costs.
- There are market opportunities for re-usable products and re-use schemes that can be exploited.
- As a consumer, buying re-usable and refillable products can save you money.

Recycling

Recycling involves the processing of waste materials to produce a usable raw material or product. This enables the same materials to be used more than once, so extending their life and maximising the value extracted from them. This has clear advantages but is not an option to be pursued at any cost. Potential disadvantages to consider include the costs of collection, transport and reprocessing, as well as the often higher costs of recycled materials.

With these factors in view:
- the Government will encourage recycling where it is economically and environmentally beneficial.

Recycling has been a feature of much economic activity for a long time and is a key factor in determining the competitiveness of UK industrial sectors. For example, the packaging industry, which currently uses 30 per cent recycled material, has a turnover of £8.5 billion and provides jobs for 150,000 people. Steel and non-ferrous scrap are two of the best-known, long-established, commercial recycling businesses.

But while the benefits of recycling are being exploited by industry, the rate of recycling of commercial and household waste is much lower. Currently, only about 5 per cent of household waste is recycled.

With this in mind, we have set targets to encourage greater recycling of household waste, including:
- to recycle or compost 25 per cent of household waste by 2000; and
- to have easily accessible recycling facilities for 80 per cent of households by the year 2000.

Why recycle your waste?
- You can use the same materials more than once, so extending their life and maximising the value extracted from them.
- You can save on the environmental and economic costs of having to produce products from scratch.
- You can save on the environmental and economic costs of disposal.
- Recycling schemes can be a good way to raise public awareness and understanding of waste issues.

Energy recovery

The best way to recover energy from waste is combustion in a waste-to-energy plant. There is substantial potential in England and Wales for the expansion of waste-to-energy

---

Re-use is a practical option in the home – for example, re-using bottles or buying refillable and repairable products

---

power generation, particularly in urban areas where it is the main alternative to disposal by landfill. We need to exploit this potential.

By recovering more energy from our waste, we will contribute to meeting our target to recover value from 40 per cent of municipal waste by 2005.

However, there are still problems to be overcome. The slow take-up of waste-to-energy in the UK in recent years reflects three main factors: the high costs of waste-to-energy compared with landfill; pollution concerns; and planning delays.

The introduction of the landfill tax from 1996 should help address the imbalance between the costs of landfill and incineration; and the *Review of Dioxin Emissions in the UK*, published by Her Majesty's Inspectorate of Pollution earlier this year, should help allay fears about pollution from incineration. In addition:
- the Government will help provide more information to local communities and local authorities about the potential role of incineration with energy recovery in a sustainable waste strategy.

Combined heat and power (CHP) schemes based on waste incineration offer even greater energy benefits than schemes which generate power only. In areas close to a large demand for heat, CHP has a particular role to play: for example, through the development of community heating schemes.

Why recover energy from waste?
- Energy recovery reduces the volume of waste for final disposal by about 90 per cent.
- It is a renewable source of energy.
- Combined heat and power schemes can increase energy efficiency by about 30 per cent.
- Energy recovery can be combined with recovery of materials for recycling.
- Incineration with energy recovery can be suitable for many difficult types of waste.

The above is an extract from *Making Waste Work: Summary*, published by the Department of the Environment and the Welsh Office.

The WEN guide to waste and how to prevent it

Information from The Women's Environmental Network (WEN)

Trees from ancient snow forests are pulped to make beer mats, newspapers and tissues. Tropical areas are devastated to mine aluminium ore for throwaway cans. Every product has a history which makes it precious. Yet, even though we are a materialistic society, we undervalue materials such as wood, paper, metal and plastics.

Our consumption habits lead to enormous quantities of materials being passed along a vast energy-intensive production line. From source, materials go via factories to supermarkets, our homes, dustbins and finally to landfill.

In contrast, nature has cycles that sustain ecosystems. One creature's waste becomes another's food. Micro-organisms turn waste into simple chemicals that plants absorb. Powered by the sun's energy, leaves grow, and the cycle starts again.

Each UK household throws away a tonne of waste per year contributing to a 20 million tonne rubbish mountain. Disposal results in long-lasting pollutants being emitted which can enter the air, soil, water and food chain.

Waste prevention involves looking at the different items that constitute waste and how they can be avoided. Reducing waste is more efficient than recycling. Recycling, like any other industrial process, produces waste itself and is never 100% effective.

The production of some materials, such as aluminium, regarded as disposable or recyclable, causes environmental degradation and is costly in terms of energy. To reduce waste, products need to be both re-usable and durable. Disposable products need to be made of minimal quantities of locally produced biodegradable materials. Using locally produced materials will reduce energy use in transportation.

To halt the unsustainable flow of materials from all round the world into our dustbins, we need a change in culture. A reduction in waste can be achieved by valuing human energy rather than products. For example washing an item, rather than replacing it with a new one, reduces waste. A shift is needed so that the systems that bring us food and other goods favour local distribution and refillable packaging. Businesses could profit by being efficient, clean and reliable rather than by promoting over packaged short-life products that are bought and discarded on impulse.

Many activities traditionally done by women, such as washing and buying food, have been devalued. This has happened alongside the growth of the consumer culture and the increasing use of overpackaged convenience products. It is no coincidence that Agenda 21, the 1992 Earth Summit Declaration, emphasises the need to value women's expertise. Women are at the forefront of change. For example, by starting dynamic small businesses such as nappy washing services time is saved and waste prevented.

Agenda 21 clearly aims for a waste-free society. People involved in some local Agenda 21 groups are realising this and setting targets. For example the Association for London Government aims for a 75% reduction in waste by 2020.

This briefing focuses on household waste. By reducing the amount and changing the things we use and throw away, we can help to save forests and avoid the pollution caused by landfill sites and incinerators in anyone's backyard.

How much waste?

Each household produces just under one tonne of waste a year. In the UK

Reasons why people recycle their waste %

People were asked about their motivations to recycle. The motivations of those who recycle are consistent across the six countries surveyed: the environment and a sense of responsibility are the major motivating factors in every country. In those countries where recycling levels are high, there are a great number of people saying recycling is easy for them to do and that it is an habitual part of everyday life

	Germany	Netherlands	Italy	Spain	GB	France
It is my/our way of helping the environment	55	54	61	48	55	54
It is what every responsible citizen should do	59	36	58	23	36	40
It is so easy it would be stupid not to	24	25	25	7	16	37
It is just a habit – I/we do not think about it much	23	39	21	17	21	38
Other	18	12	5	10	16	17

N.B. These tables add up to more than 100%, as the respondents were allowed to give multiple answers.

Source: The Aylesford Newsprint Recycling Report

we discard nearly one million tonnes of plastic, two million tonnes of glass and over six million tonnes of paper every year in household waste.

What happens to our waste?
Landfill
The majority of our rubbish, approximately 90%, is taken to landfill sites. Landfill rubbish generates the potentially explosive methane gas that contributes to global warming, and leachate, a liquid formed from the breakdown of rubbish, which can enter and pollute groundwater.

Incineration
About 5% of our rubbish is incinerated. Incineration creates toxic substances such as dioxins in air emissions and ash. 30% of the original tonnage still has to be landfilled as ash, in which heavy metals from the waste become concentrated. A new incinerator can cost £40 million and operators demand 25-year waste supply contracts from local councils.

Waste can be burnt to generate electricity. 'Waste to energy' incinerators receive a renewable energy subsidy from our electricity bills. However, energy has already been lost in the manufacture of throw-away items, and generating electricity from burning waste is in itself inefficient.

Recycled
Only 5% of our waste is recycled. The Government aims to increase this to 25% by 2000. While recycling is important, it only deals with waste after it has been produced. In an ideal world it would be most appropriately used for re-used and worn-out objects.

Waste and the economy
We pay for our rubbish to be created, collected, transported and disposed of. It costs £750 million a year to pay for the new materials needed to make the things we throw away and £520 million a year to dispose of them.

Hazardous waste
Many of the products we use in our homes and gardens contain toxic chemicals and metals that are routinely disposed of down the sink or via our dustbins to landfill. Examples include batteries, pharmaceuticals, some cleaning fluids, smoke alarms, fluorescent tubes, oil and oil filters, paints and garden chemicals.

As these products cause persistent pollution the best option is not to use them.

Non-toxic alternatives for paints and cleaning products are available and rechargeable batteries reduce disposal. Toxic substances that can't be avoided at present can be disposed of more safely via the retailer or the council.

What is in your bin?
Paper	32%
Other	15%
Compostable	22%
Plastics	10%
Glass	8%
Iron & steel	6%
Disposable nappies	4%
Textiles	2%
Other Metals	1%

Packaging
- 30% of domestic waste is packaging.
- It is estimated that we spend £10 on packaging in every £75 we spend on groceries.
- The UK packaging industry has an annual turnover of £11 billion and consumes 5% of the total energy consumption and vast quantities of natural resources.

The effects of waste on our health and the environment
Dioxin is one of the most toxic substances known. It is created by the incineration of waste as well as the manufacture of paper and metal. When waste contains chlorine in the form of PVC plastics and chlorine-bleached paper, higher dioxin levels result. Dioxins in the air eventually reach vegetation and enter the food chain. Foods like meat, fish and dairy products contain higher levels of dioxins which become concentrated in fat. The main effects of dioxin are impaired immunity to disease, reproductive disorders and cancer.

Some plastic film and printed paper used in packaging contains phthalates which migrate into food, particularly fatty foods. Phthalates are hormone disruptors which mimic the body's natural oestrogen. These oestrogen mimics may contribute to breast cancer and male reproductive disorders. Other types of oestrogen mimics are released from the linings of food cans. The 50% decrease in the male sperm count over the last forty or so years could be a result of environmental pollutants such as phthalates and dioxins.

Heavy metals such as cadmium, lead and mercury are present in a variety of products such as batteries, fluorescent tubes and energy-efficient light bulbs. These metals do not degrade and if they enter the food chain they can cause liver, kidney and brain damage in humans.

Live life – don't waste it!
Change in our attitude to materials is essential and inevitable. By valuing materials more we will prevent them becoming waste by re-using them. We will also use less. Homes will be less cluttered, but what is in them will be more valued and long lasting. There will be an imaginative shift in lifestyle, not to frugality, but involving more ceremony, fun and enterprise. Waste prevention should be seen as a positive way of meeting needs that values both human ingenuity and materials.

Refills
It is cheaper and greener to wash, sterilise and re-use containers than to produce new glass, paper or plastic ones. Refilling glass bottles uses 75% less energy than making them. Refill systems favour local distribution and if the trippage rate (number of times refilled) is high and the transport distance short, waste is reduced to a minimum. Ecover, which has recently introduced a refill system, estimates that this reduces the cost of the product by 15 to 20%. The Body Shop offers a discount of 20 to 50p on each refill.

A deposit system was introduced in Denmark when disposable drink bottles were banned in 1979. Over 150 million bottles are returned every year and refilled on average 32 times. In the UK only 19% of soft drinks come in returnable bottles. Milk delivery is the most widely used refill service supplying over 12 million households. Supermarket competition has led to a fall in milk delivery services from 90% of all household milk in 1970 to 45% in 1995.

Re-usables

There are re-usable alternatives to one-use disposable products such as bags, napkins, nappies and sanitary towels.

Nappies

Over nine million nappies are thrown in the bin every day, making up 4% of household waste. Nappy manufacture is a highly polluting process using trees from as far away as Canada and the plastic coating prevents them from decaying for decades in landfill. Fitted re-usables and terries are available and by using a nappy washing service re-usables can be convenient too.

Sanitary protection

Every year over two billion sanitary towels and tampons are flushed down the toilet, causing visible pollution and harming marine life. Bagging and binning is less damaging, but disposal is avoided by a growing movement of women who are using re-usable forms of sanitary protection.

Alternative shopping

Group buying

Many whole-food shops will offer a bulk buying service for groups of people who order their food together. All you need is weighing scales, storage space and lots of refillable containers. Buying wholesale can save up to 35% in price and packaging is reduced.

Organic veg delivery services

As well as reforging links between communities and local farms the delivery of organic vegetables in durable re-usable boxes with no wrapping reduces waste to a minimum.

Local exchange trading systems (LETS)

Local exchange schemes enable people to draw from a whole wealth of local skills and resources. A local credit unit is used instead of money.

Composting

Composting is the process of speeding up the breakdown of biodegradable kitchen and garden waste to create organic fertiliser. One-third of house-hold waste including vegetable peelings, tea leaves, egg shells, shredded paper and card is compostable.

This recycling of waste eliminates the need for chemical fertilisers and reduces the formation of meth-ane gas in landfill sites. It is easy to establish a composter in the garden and 'worm-bins' enable waste to be composted indoors.

© *The Women's Environmental Network*
February, 1996

One week's waste from a family of two adults and two pre-teen children

Waste	kg	Items
newspapers/ magazines	3.6	national daily and local papers, weekly free papers, radio & TV guide, magazines, comics & holiday brochures.
Other paper/ board	0.6	cream cracker packet liner, 2 egg cartons, 3 orange juice cartons, butter wrapping, toothpaste carton, junk mail, cake box, used envelopes, 3 toilet roll tubes, 2 cereal boxes, pizza box, soap wrapper.
Glass	1.6	squash bottle, 2 jam jars, 2 wine bottles, whisky bottle, peanut butter jar, instant coffee jar.
Ferrous metal	0.9	5 tomato cans, 4 soup cans, 2 baked bean cans, sweetcorn can, 2 drink cans, tinned carrots can, deodorant aerosol can, 2 jam jar lids.
Aluminium	0.3	2 sardine cans, 6 drink cans, toothpaste tube, whisky bottle cap.
Plastics	0.6	12 yoghurt pots, soft fruit punnet, 6 two-pint milk bottles, shampoo bottle, 2 litre fizzy drink bottles, 2 carrier bags, 3 pre-packed cheese wrappers, margarine tub, bacon packet, meat tray & plastic wrap, ice cream tub, frozen peas packet, aerosol lid.
Putrescibles	4.5	fruit & vegetable peelings, tea bags, eggshells, chicken carcass, bacon rind, stale bread crusts.
Other	1.8	2 pair tights, broken mug, kitchen sponge, 2 wine corks, calculator button battery, can of old engine oil.
TOTAL	**13.9**	

• The above is an extract from an information pack called *Wise up to Waste*, published by Waste Watch. See page 39 for address details.

© *Waste Watch*

The law against littering

- Did you know it's a criminal offence to drop litter?

- Do you know what the maximum fine is and who can prosecute?

- Can you be fined for throwing litter from a car?

- What powers do litter wardens have?

This article explains what the law is about dropping litter.

The offence of littering

People who carelessly discard litter in a public open place are committing a criminal offence. (Except where it is allowed by law or done with the consent of the owner or occupier of the land.)

Prosecutions for littering are brought under section 87 – Offence of Leaving Litter – of the Environmental Protection Act 1990. The offence is:

'If anyone throws down, drops or otherwise deposits in, into or from any place to which this section applies, and leaves any thing whatsoever in such circumstances as to cause, or contribute to, or tend to lead to, the defacement by litter of any place to which this section applies, he shall . . . be guilty of an offence.'

There were about 1,000 police prosecutions for littering in 1995 (England and Wales). The average fine is around £100 plus any costs the court awards. However, a person convicted of this offence could be liable to a maximum fine of £2,500 (a level 4 offence on the standard scale).

What is litter?

The term 'litter' has a very wide interpretation. As well as wrappers, cans, bottles or packaging, 'any thing' includes food, small items such as cigarette ends or large items such as bags of rubbish. The offence relates to what is *done* with the litter, rather than what it is.

Natural matter such as overgrown vegetation, weeds or leaves which have fallen from trees are not classed as litter.

Dog faeces are classed as 'refuse', not litter, and are subject to different laws.

Prosecutions

Prosecutions are usually undertaken by the police, through the Crown Prosecution Service, but local authorities can also prosecute. Cases are heard in the local Magistrate's Court.

If you witness a littering offence you can report it to the police or chief solicitor at your local authority and they will decide if legal action is appropriate. You should gather as much evidence as possible to support your case, such as the time, precise location, a description of the litter and identity of the litterer. If the case goes to court and the accused pleads not guilty you may be called upon to give evidence. However, it is common for the accused to plead guilty by letter and for the case to be dealt with in his/her absence. As an alternative to prosecution, the police or the local authority may warn or caution the litterer.

It is possible for the citizen to pursue a private prosecution. However, you will need strong evidence to prove your case, preferably with a witness. Legal aid will not be available. We recommend you seek legal advice before proceeding.

Litter thrown from vehicles

Throwing litter from vehicles onto public land is also a littering offence under section 87 of the Environmental Protection Act 1990.

If the registration number is taken the registered keeper's identity can be obtained from the Driver Vehicle and Licensing Agency by the police or the local authority. Witnesses should note whether the driver or passenger threw the litter and take a good description of the vehicle and offender in case it is needed in evidence. Offences can be reported to the police or local authority and they will decide whether to take further action.

Litter wardens

As an alternative to taking litterers to court, a local authority can appoint officers to issue £25 fixed penalty notices to litterers. This has the benefit of dealing with littering as and where it happens. The offender is given 14 days in which to pay the fine. If it is not paid he/she can be taken to court under section 87 of the Environmental Protection Act 1990 and fined up to £2,500.

Disposing of litter:
Put litter in a bin
If there are no bins
TAKE YOUR LITTER HOME

© Tidy Britain Group

Beachwatch 1996

Information from the Marine Conservation Society

Summary

Beachwatch is a practical conservation initiative, organised by the Marine Conservation Society with sponsorship from Reader's Digest. The Beachwatch campaign was launched in 1993 to raise awareness of the issue of marine debris, monitor litter on Britain's beaches and encourage action to reduce marine pollution at source.

The Beachwatch '96 beach clean and survey took place on 21 and 22 September 1996, to tie in with the International Coastal Cleanup occurring in over 70 countries worldwide. A total of 237 beaches in England, Scotland, Wales, Northern Ireland, Eire, Malta and the Channel Islands were cleaned by 3,371 volunteers. In the UK 203 of these beaches covering 196 km of UK coastline were surveyed in detail. In total, 291,228 items of debris were recorded, weighing an estimated 17,518 kg (17 metric tonnes).

The Beachwatch '96 results show a 10% decrease in the number of items recorded per kilometre surveyed as compared to Beachwatch '95. 1,482 items per km were recorded in Beachwatch '96 compared with 1,635 items recorded per km surveyed in Beachwatch '95; 967 per km surveyed in Beachwatch '94; and 572 per km surveyed in Beachwatch '93.

Plastic, the most persistent debris material, has consistently amounted to over 50% of all debris recorded ('96-52%; '95-53%; '94-54%; '93-52%). Sewage-related debris (SRD) comprised 13.5% of the total, the highest proportion recorded since 1993, but this can be attributed to an increase in cotton bud sticks recorded ('96-154/km; '95-92/km; '94-75/km). The quantity of sewage-related debris other than cotton bud sticks recorded per km surveyed has decreased from previous years ('96-46/km; '95-107/km; '94-62/km). Cotton bud sticks made up over 75% of the SRD items recorded. The most common 12 items of debris made up 69% of the total item count, with just six item types alone representing 48% of the total items recorded (plastic pieces, cotton bud sticks, plastic rope & cord, plastic crisp & sweet wrappers, plastic caps & lids and plastic bottles).

Four sources were identified as major contributors to the amount of litter on UK beaches: direct littering by recreational users (22.1%), shipping (17.4%), sewage outfalls (13.5%) and fishing vessels (12.5%). The proportional contribution of these four sources is consistent with previous Beachwatch surveys.

Tourist/recreational litter was the most common source of litter on beaches in Northern Ireland (54.2%), the Channel Islands (28.3%), Wales (23.5%) and England (21.1%). Sewage-related debris was the most common source of litter on beaches in Scotland (34.5%) and the Isle of Man (28.4%).

Recommendations

Reduction of pollution from ships and other vessels:

- Mandatory development of comprehensive and adequate port and ship waste management plans.

Beachwatch '96 Summary Statistics	
Volunteers	3,371
No. of bags	2,438
Weight of rubbish (kg)	17,518.6
No. of beaches surveyed	203
Length surveyed (km)	196.5
Area surveyed (km)	2,419.8
Total items surveyed	291,228
Average items/km	1,481.7

- Improvement in the scope, use and operation of port waste reception facilities by making provision a statutory duty.
- Introduction and use of Garbage Record Books by law.
- Introduction of minimum penalties for illegal dumping of wastes as a deterrent to polluters.
- Implementation of a public register of legal notices, offences and fines relating to pollution of the marine environment from shipping.
- Further educational material for ship owners, ship operators, port users, fishermen, recreational boat users and all other interested parties with regard to their responsibility in preventing marine pollution.
- Establishment of Marine Ecological High Risk Areas (MEHRAs) as recommended in the Donaldson Report.
- Encouragement of countries which have not ratified or implemented MARPOL through their own legislation, to do so.

Reduction of the input of sewage-related debris:

- Secondary treatment of sewage as a minimum for all coastal sewage outfalls.
- UV disinfection for all sewage discharges affecting bathing waters.
- Immediate implementation to improvements to the EC Bathing Water Directive.
- Designation of more bathing waters – expanding the term of cover all recreational waters.
- Full implementation of the Urban Waste Water Treatment Directive.
- Labelling of sanitary products with the correct disposal information.

- Minimisation of the environmental impact of sanitary protection, by the reduction or replacement of plastic materials with biodegradable (non-plastic) alternatives.
- Provision of disposal facilities for sanitary waste in all public toilets.
- Provision of information and education to encourage the public to *'Bag it and Bin it. Please Don't Flush It.'*
- Education in schools on the issue of sanitary waste.

Reduction in inputs of plastics:
- Provision of adequate facilities and services for the public to dispose of their rubbish, along with publicity on their location etc.
- Incorporation of beaches in local authority waste management plans.
- Encouragement of voluntary local initiatives such as Adopt-A-Beach by local authorities.
- Improvement of the combined sewer overflow system by water authorities.
- Encouragement of the public to recycle and buy items with less packaging.
- Expansion of national plastics recycling schemes by Government.
- Enforcement of anti-littering laws to deter fly-tipping and littering by the public.
- Implementation of plans to minimise packaging waste by Government and industry.
- Incorporation into school curricula of educational material on the sources and effects of marine debris and ways of reducing the problem at source.

Individual responsibility:
- Individuals must dispose of their waste in the most environmentally sensitive way:
  Reduce – Re-use – Recycle.

• The above is an extract from *1996 Nationwide Beach-clean & Survey Report*, published by the Marine Conservation Society, sponsored by Reader's Digest.

Tourists blamed for bulk of beach debris

Tourists are the biggest beach polluters, accounting for more than a fifth of the rubbish collected from seaside resorts, it was disclosed today. A huge beach clean-up operation collected 17.5 tonnes of rubbish from 203 stretches of Britain's coastline. Items found included an entire railway sleeper, a fridge, a washing machine, a Russian salt-pot and an aerosol can from Israel. Holidaymakers and other visitors were responsible for dropping more than 18,000 crisp and sweet packets, more than 8,000 drink cans and nearly 10,000 cigarette ends.

But the Beachwatch '96 clean-up, organised by the Reader's Digest and the Marine Conservation Society, did show that beaches were becoming slightly less polluted.

An average of 1,482 items of debris per kilometre were gathered from the 203 beaches, compared with 1,636 per km from 196 coastal stretches in the previous clean-up.

After tourism, shipping was the next worst polluter – blamed for 17.4 per cent of debris, including nearly 27,000 lengths of rope and cord – followed by sewage-related debris, including more than 30,000 cotton buds and 6,000 sanitary products. The 3,300 clean-up volunteers also collected more than 4,500 lengths of fishing line that can entangle and kill wildlife.

A spokesman for the Marine Conservation Society said: 'We have to change attitudes towards litter and encourage individual responsibility in today's throwaway society.'

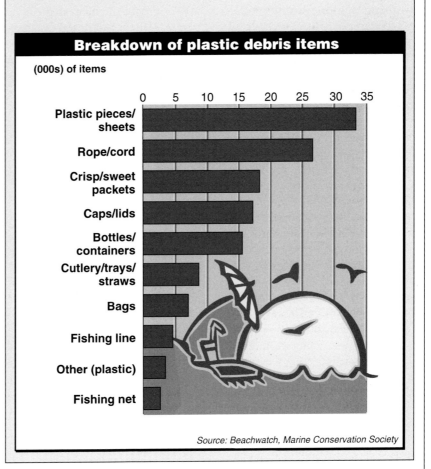

Breakdown of plastic debris items

(000s) of items

- Plastic pieces/sheets
- Rope/cord
- Crisp/sweet packets
- Caps/lids
- Bottles/containers
- Cutlery/trays/straws
- Bags
- Fishing line
- Other (plastic)
- Fishing net

Source: Beachwatch, Marine Conservation Society

Looking at litter and recycling

In the natural order of things, recycling – making new things from old discarded ones – is a fact of life, and nature's 'litter' usually has built-in recyclability. Take trees for example. Many trees drop all their leaves in autumn, but evergreens drop some leaves all year round. The discarded leaves on the ground form a mixture of twigs and leaves called leaf litter. The leaf litter rots down to become part of the soil and forms a natural soil enricher, supporting the healthy growth of new vegetation.

Unfortunately our own litter is not so obliging, and a growing proportion of the things we discard do not rot away or 'biodegrade' but stay around indefinitely. In their manufacture, many use up the earth's valuable resources and produce harmful waste products, yet have only a limited useful life.

Lethal litter

Some items are not only unsightly as litter, but are also dangerous to wildlife.

- Yoghurt cartons can trap inquisitive hedgehogs who move into the carton attracted by the remains, but then – because of their prickles – cannot move backwards to get out.
- In waterways and the sea, discarded plastic loops which originally held drinks cans together can catch and kill diving birds and fish. On land, the loops can trap hedgehogs and other small mammals. One way to avoid this is to cut the plastic loops before they are discarded.
- Floating plastic bags look like appetising food to marine life, yet can kill if they are swallowed.
- Discarded fishing lines can maim and kill waterbirds.

Your local wildlife trust may help organise 'clean-ups' of certain areas, for instance ponds and streams where litter is a particular threat to wildlife. Joining a wildlife trust means you could take part in practical action.

Using old aluminium from used cans is a great energy saver – for every eight cans recycled, a litre of fuel is saved

---

Even putting rubbish in the right place – in bins, can cause problems. After it's collected, it is taken to places called landfill sites – large natural or manmade holes in the ground. But we're running out of these convenient 'holes' and, in addition, the liquid from these heaps of rubbish can soak into the soil and can be dangerous.

Recycling makes sense

Many local authorities now have a Recycling Officer and provide places where unwanted items can be collected for recycling. Melting old bottles to make new glass uses less heat energy than making totally new glass – about one litre of oil is saved for every four bottles you put in the bottle bank.

Here are some examples of sensible recycling:

- Give clothes, toys and books to charity shops or jumble sales.
- Use worn-out clothes as cleaning rags.
- Buy sticky labels from your favourite charity and re-use envelopes.

- Advertise things like washing machines, freezers, fridges and furniture in your local paper – someone else can probably use them. People sometimes dump freezers and fridges on waste ground. This is potentially dangerous, not only because young children and animals can become trapped inside, but also because fridges and freezers contain CFCs. These are substances which originally helped the fridge or freezer keep food cool, but which are also an environmental hazard. If you can't sell unwanted household appliances, contact your local council for how to dispose of them safely.
- Make compost from kitchen or garden waste and give food scraps to birds.
- Take used car oil to a garage for safe disposal by recycling.

- Get together as a group to collect paper – merchants are only interested in large quantities. You may need a fire-safe place to store it. The British Waste Paper Association can advise you.

Can recycling

Drinks cans are made from aluminium or tin plate steel (steel covered with a thin coat of tin). Making cans is very expensive and uses a great deal of energy. Using old aluminium from used cans is a great energy saver – for every eight cans recycled, a litre of fuel is saved. If you squash the cans before taking them to a collection point, they take up less space. Fewer lorry trips are then needed to collect and empty the bank, which saves even more energy.

Litter and the law

In 1990 a new law called the Environmental Protection Act made sure that schools and colleges, local authorities and government departments had a duty to keep their land free of litter.

Local authorities have been given new powers to help them fight litter.

- They can insist that shops and take-away restaurants keep their street clean.
- They can appoint litter wardens who can give people a £10 fine for dropping litter if they do not pick it up straight away.
- The maximum fine for littering is now £1000.

Did you know . . .

Each year in Britain, we throw away and bury cans worth a total of £24 million – about 90 drinks cans per person.

New UK Government strategy on recycling

The Department of the Environment has published the Government's strategy for sustainable waste management in England and Wales: *Making Waste Work*. In it they have set targets, some for the first time, for waste control, recovery and recycling. Environment Secretary John Gummer wants industry, business, local and central government, as well as householders, to play a key role in reducing, re-using and recycling their waste. At the launch, held at Cremorne Wharf Recycling Centre in London, Mr Gummer stated: 'The strategy aims to improve the way we manage waste . . . This message is aimed at everyone.'

The main targets set in the new strategy include:
- the reduction of controlled waste going to landfill to 60% by 2005.
- the recovery of 40% of municipal waste by 2005.

- 40% of houses with gardens to carry out home composting by the year 2000.
- the availability of easily accessible recycling facilities for 80% of households by the year 2000.

The Government believes that these targets are ambitious but achievable. The new policy framework provided by this new strategy builds on the ideas in the Government's Sustainable Development Strategy published in January 1994 to ensure economic development along with the protection and enhancement of the environment. To tie in with this, a seminar hosted by John Gummer and the President of the Board of Trade was held on 2 April 1996.

Feedback to *Making Waste Work* has been varied: Friends of the Earth are highly critical, complaining that: 'This has been a wasted opportunity to address the urgent problems of waste. Whilst the Government has started to acknowledge the extent of the problem, the lack of waste reduction targets and any statutory backing means that their initiatives are likely to fail to meet our existing European commitments.'

They also condemned the new strategy for promoting incineration with energy recovery to the same level of priority as waste re-use and recycling. On the other hand, the Environmental Services Association (ESA) welcomed the White Paper. Their chairman, Peter Neill, commented: 'ESA agrees with Government that landfill will provide the best practicable environmental option for the majority of wastes for the foreseeable future.'

The marine debris scandal

Litter in the ocean and on our beaches is one of the most visual reminders of humankind's ever-increasing impact on the marine environment

It is becoming difficult to find any small piece of coastline that does not bear the mark of human existence – bottles, sanitary towels, cigarette ends, plastics and fishing net. Our debris not only looks terrible, but also threatens wildlife, human health and the local economies which depend upon the coast and sea.

Where does it come from?
There are various ways that debris can enter the marine environment. The most obvious is direct littering by beach users. Other sources include sewage, dumping by ships and the rubbish carried downstream in rivers. The debris can be divided into nine main types – plastics, polystyrene, rubber, wood, metal, sewage-related debris (SRD), paper, cloth and glass.

The problem with plastics
Plastic is the most common man-made material found at sea and on beaches world-wide. Being both durable and versatile, it has become extremely popular in the consumer and fishing industries. However, the durability is a problem once plastics enter the environment as it can take 400 years for one piece to completely break down. Marine animals often become entangled in plastic or mistake it for food – this can lead to starvation and even death.

Wildlife entanglement
Wildlife is severely affected by debris. Of the 115 species of marine mammal, 47 are known to become entangled in and/or ingest marine debris. Numerous birds and turtles are also affected.

Entanglement occurs when debris, such as fishing net or plastic strapping bands, becomes tangled around the body or limb of an animal. Young seals and cetaceans are particularly prone to this as they are

By Wendy Higgins

naturally inquisitive and can become snared during play.

Large debris can trap animals beneath the surface of the sea, causing air-breathing individuals to drown. Once trapped, the animal also becomes vulnerable to death by starvation or predation. Entanglement by smaller debris can dangerously increase the 'drag' of an animal through the water. Consequently, the animal needs to use more energy and may have its ability to forage and escape predators severely impaired. Small debris, such as drinks-can yolks, can be equally dangerous to animals and birds, becoming so tightly bound as to cut off limb circulation.

Debris is also incorporated into nesting materials by seabirds. Entanglement of young chicks can lead to life-threatening injuries and debilitating deformities. Plastics used for fishing nets, strapping bands and so on are the greatest threat – once caught the animal is unlikely to be able to release itself. One report suggests that over 30,000 seals die every year of entanglement, usually in discarded fishing nets.

Ingestion
Ingestion of marine debris can also be a wildlife killer. Turtles often mistake floating plastic bags for their favourite food – jelly fish. Larger mammals, such as sperm whales and sharks, can make the same mistake with plastic sheeting. The accum-

---

New agreement achieved on waste dumping at sea

A new global agreement was reached today, by the 75 member countries of the London Convention 1972, to strengthen controls on the dumping of wastes at sea. The agreement prevents any waste material being incinerated at sea and means that new rules will strictly limit the type of waste which may be dumped at sea.

The two-week Special Meeting of the London Convention, which aims to protect the global marine environment, ended today. Central to the agreement is a new initiative to examine all other waste options before dumping at sea is undertaken. The agreement stresses the importance of science in understanding and addressing marine pollution, and seeks effective scientific and technical co-operation between countries.

In welcoming the agreement, the Minister Douglas Hogg said today: 'The London Convention 1972 has been a major achievement in that it has protected the world's seas from unregulated dumping of waste over the past twenty years. I believe today's agreement is a significant development which will further strengthen controls on dumping and provide a basis for the continued protection of the marine environment well into the next century. The provisions designed to enhance scientific and technical co-operation between countries will contribute towards our objective of encouraging the widest possible membership of the new agreement.'

© MAFF
November, 1996

ulation of plastics in the animals' bodies can be lethal. Ingestion can lead to a range of physical disabilities. The consumption of sufficient small plastic pieces or 'micro-plastics' makes an animal feel full, thus interfering with nutrient intake.

Ingestion can result in further problems, including scarring and ulceration of the digestive tract, starvation due to blockage of the food passage or poisoning by toxic chemical pollutants released by debris.

Sewage-related debris

Sewage-related debris (SRD) such as tampons, sanitary towels and nappies is a particularly unsightly problem on our beaches. It also presents a potential health risk to both humans and wildlife. Incidence of SRD on the beach is an accurate indicator of the general state of the nearby ocean.

Every day, over 300 million gallons of raw or partially treated sewage are discharged into UK seas – that's around 2–3 buckets per person! Other solids reach our seas via the lavatory too because of a tendency to use the toilet as a 'wet bin'. 66% of women in Britain flush their sanitary protection (SANPRO) away, that's a massive 2 billion sanitary items each year. Disposal of such items down the toilet causes blockages in the sewerage system which can lead to overflows directly into rivers and seas.

Risk to human health

The risks to humans, animals and the environment from SRD are obvious. Contaminated seas often contain a high degree of pathogens (disease-causing micro-organisms), whilst the bacteria that thrive on SANPRO can survive in sea water for weeks. Raw sewage and SRD also constitute a health risk to swimmers, surfers and divers, who suffer from ear, nose and throat infections and gastro-enteritis.

Extra nutrients

The fragile balance of the marine environment is vulnerable to nutrient enrichment caused by sewage. Sewage is a highly effective fertiliser and when deposited in a small area

can cause certain types of algae (phyto-plankton) to multiply very quickly. These algae can be toxic in high concentrations, block out natural light and use a large proportion of the available oxygen in the water, thus smothering other organisms.

Looking to the future

If the current situation is to be improved, the issue of marine debris must be taken seriously at every level. Government and industry must take the lead in changing attitudes and actions. There is a need for tougher measures to counteract pollution, plus better enforcement of existing legislation. Since the setting up of MARPOL*, the UK Government has consistently failed to provide adequate, cheap and convenient port waste facilities. These are essential if fishing vessels, tankers and other craft are to be deterred from illegally dumping their waste at sea. However, despite the Government's apparent commitment to MARPOL, it is clear that enforcement has so far failed. There has been only one prosecution in the UK for illegal dumping of

waste, resulting in a fishing vessel being fined just £100. As Lord Donaldson stated in his report following the Braer oil tanker disaster, 'Inadequate facilities are a positive invitation to ignore the rules.'

It is equally essential that we see an improvement in the UK of our sewage treatment procedures. Currently, approximately 40% of major UK coastal sewage outfalls are not required to have even primary screening to remove large debris items. The Government could introduce legislation to make secondary treatment the minimum national standard to be achieved at all outfalls. Individual people should be encouraged to dispose of sanitary waste responsibly. We have a major role to play in preventing as well as creating marine pollution, and it is through personal action that many marine debris problems can be resolved.

* United Nations' International Convention on the Prevention of Pollution from Ships, 1973.

© The Living World, Earthkind
Winter, 1996

Healthy seas are vital

- Over 70% of the world's surface is covered by seas and oceans
- The oceans regulate the global climate
- Algae in the oceans produce over half of the oxygen that we breathe
- The sea provides the sole source of protein for one-third of the world's population

and yet:

- Every day, ships throw over 5 million items of waste into the sea
- Every day, over 300 million gallons of sewage are discharged into UK coastal waters
- Every day, there are still up to 30,000 miles of drift gill nets set, killing everything in their path, including more than 1,000 dolphins and 3,000 sharks – every day!
- Every year, three times as much rubbish is dumped into the world's oceans as the weight of fish caught
- Every year, an estimated one million seabirds and 100,000 marine mammals and sea turtles suffer cruel deaths from entanglement in, or ingestion of, plastics
- Every year, over 100 million sharks are killed, many solely for their fins
- Over 60% of the world's coral reefs are threatened as a result of sedimentation, pollution, nutrient enrichment and harvesting for the marine curio and aquarium trade and bleaching due to rising water temperatures as a result of global warming.

© Marine Conservation Society

Recycling – the facts

Information from the Pulp & Paper Information Centre

Recycling is one of those rare activities from which practically everyone derives a benefit. The paper industry is a recycler 'par excellence' – and has been for hundreds of years. Before woodpulp became his major raw material, the papermaker used old rags. Today, although some rags are still used, by far the largest material for recycling is paper and board itself. Papermakers are therefore understandably aggrieved when, as a result of widespread misunderstandings in the public mind, custard pies in the face rather than bouquets in the hand tend to be the order of the day. The misconceptions are not limited to the industry's record in recycling: its use of wood is the basis of, if anything, even more confused thinking and public mis-statement.

The following are some of the most commonly asked questions received by the Pulp & Paper Information Centre.

Q. Is the paper industry instrumental in destroying the Brazilian rainforests?
A. No. Quite apart from any other consideration, trees growing in the rainforests (mixed tropical hardwoods) are not a good source of papermaking fibres. Trees for papermaking, mainly eucalyptus but softwoods too, are grown in the coastal areas (not in the rainforest areas) of Brazil where the climate is ideal for rapid growth. Destruction of the rainforests is caused by the demand for certain decorative slow-growing hardwoods and (mainly) by the short-sighted policy of clearing land for farming.

Q. Does recycling waste paper save trees?
A. This is one of the major recycling 'myths'. Only 10% of the industry's fibre needs are derived from UK timber and these come mainly from thinnings which have to be taken from the forests in order that the remaining trees can grow to healthy maturity. Wood is used in four mills in the UK and converted into mechanical pulp or semi-chemical from mixed hardwood – such as birch, poplar and the tops of beech or oak felled for other purposes. Forget the concept of 'saving' a lovely beech tree or a mighty oak; anything over 16 feet in height is not wanted primarily for papermaking. Industrial wood comes from commercial forests where the trees have to be harvested at the correct stage for the purpose in mind. For sawlogs (going to construction, furniture, etc.) mature trees are indeed required and then papermakers stand by to take the otherwise waste residue.

Recycling therefore does not 'save' trees but it does ensure that the life of the fibres taken from trees is prolonged.

Q. Do we need to recycle waste paper in order to conserve the world's limited tree resources?
A. No. Trees for industrial use are grown as a crop just like cereals and are merely harvested at longer intervals. Modern silviculture ensures greater efficiency than ever before in cultivating trees and more wood is growing at all times than is being used. Wood is believed to be the only industry raw material which is infinitely renewable. The world's paper industries recycle a lot of waste paper. The fact that they do obviously makes it easier for the world's commercial tree growers to maintain a potential supply greater than current or foreseen demand. There are more important and direct reasons for recycling paper than 'saving trees'.

Q. What are they?

A. i) Waste paper supplies are an economic necessity to half the paper and board industry in the UK. Without it, many mills could not compete with imports from countries where cheap wood is plentiful and paper, can be made in integrated plants with trees or chips coming in at one end and paper coming out at the other.

ii) The availability of waste thus enables products to be turned out in the UK instead of being imported, thereby contributing positively to the balance of payments and creating employment not only in the paper industry but also in the waste collecting and processing industries.

iii) Waste materials are usually buried or burnt and the cost of such disposal can be very high.

Q. If recycling is so advantageous then why don't we use even more waste paper?

A. There are many reasons including:

i) Some papers simply cannot be made from waste, eg. bank note paper, filter papers and cigarette paper.

ii) Others may be made with a small percentage of waste but the waste has to be unprinted and clean – virtually the same material as the pulp it would replace, the only difference being that it has been through a paper machine. Such waste is often scarce and represents only a tiny proportion of what is collected.

iii) Other papers may be made from high-grade clean waste on which there is print or writing. This too is scarce since although large quantities arise they tend to be mixed with other waste and outsorting is expensive. Office recycling schemes where sorting takes place at source should therefore be encouraged.

iv) To make collection worthwhile there has to be a demand, i.e. there must be paper machines capable of converting the waste into fresh paper and board and the resultant output must be saleable.

v) Most waste paper is used for packaging papers and boards, newsprint and tissues. There have been major advances in the use of waste in these sectors – but the machines have to be there before the waste is collected.

vi) The UK industry already utilises waste for more than half its fibre needs. Further progress will undoubtedly be made; a major newsprint mill based entirely on waste has been announced and plans are already in hand with active government and local authority support to bring out more old newspapers and magazines from households for the purpose. Progress from such a high level of existing usage is difficult.

Q. So where do waste-based printings and writings fit in? Why aren't they more generally available?

A. Sections ii) and iii) of the previous answer apply. You cannot take a mixed batch of paper from the general garbage and reprocess it into clean white paper suitable for letterheads and six-colour printing – yet. What can be done, however, is to produce printings and writings papers for a variety of purposes from a variety of high-grade sorted waste papers.

© Pulp & Paper Information Centre

EC Directive on Packaging and Packaging Waste

After many months of detailed negotiations the EC Directive on Packaging and Packaging Waste came into force on 31 December 1994 to set the targets for recovery (recycling plus energy recovery by incineration) and recycling (turning the waste into useful, saleable products). The agreed targets for each Member State are:

- By 30 June 2001, between 50% and 65% by weight of packaging waste will be recovered, and
- within this target, between 25% and 45% by weight of packaging waste will be recycled, with a minimum of 15% by weight for each material (glass, aluminium, etc.).

Duales System Deutschland

Germany was one of the first EC countries to plan action on its waste problems. The German government planned to tackle the problem through, amongst other things, putting a deposit on all packaging items so that customers would return their used packets, bottles and jars for a refund.

This scheme was not brought in because the German packaging industry offered its own solution, which would avoid compulsory deposit legislation which it saw as unworkable. The resulting scheme is called Duales System Deutschland (DSD). It is a totally separate waste packaging collection scheme funded by over 400 German companies which identify their packs with a round green symbol of two curved arrows. All waste having this symbol is collected either in yellow second dustbins given to households or through packaging collection banks provided at a density of one for every 500 people.

The companies which operate the collection schemes have to sort, bale and send the materials to businesses which can recycle the packaging. The baled packaging can be given away to anyone who will use it because the cost of the scheme is covered by a charge on all the packages which carry the green symbol. The costs are passed back to the consumer and are thought to be between £2 and £4 per household per week.

• The above is an extract from an information pack called *Wise up to Waste* published by Waste Watch. See page 39 for address details.

© Waste Watch

The green shopper 'is alive and well'

By Charles Clover, Environment Editor

Claims by supermarkets that the green shopping revolution has ground to a halt and demand for environmentally friendly goods is waning are wrong, according to a study by the National Consumer Council.

Its report, published today, says the green consumer is alive and well, representing a third of all consumers. But he or she feels ill-informed by the plethora of confusing labels and wants cheaper, better-recognised products, with more meaningful, authoritative descriptions as to why they are environmentally friendly.

The council argues that consumers are looking for 'green' features from recognised makes as these are both more effective than out-and-out green lines and better value for money. They are also more likely to be on special offer.

The report identifies five types of consumer:
Affluent greens and **young greens**, who together make up 36 per cent of the population, are committed to green consumerism. Affluent greens – and those young greens who can afford to do so – act on their beliefs wherever they can find the products they need. They were generally comfortably off, in higher-status jobs, middle-class, more likely to be female and aged 25 to 50.

> *The green consumer is alive and well, representing a third of all consumers*

Young greens, at around 17 per cent of the market, were mostly students or unemployed. They tended to do a fair amount of recycling and to buy green but had limited market power.

Recyclers and **careful spenders** make up another 38 per cent of the population and act in an environmentally friendly way, although they do not usually buy green.

Recyclers regularly take their cans, newspapers or bottles to banks and save their kitchen waste for compost.

Careful spenders, four in 10 of the population and more likely to be in clerical or skilled manual occupations, say they reduce their consumption of household fuels, petrol and water for environmental reasons.

Sceptics
Sceptics (26 per cent of the population) are determined not to buy green under any circumstances. The profile of a sceptic was an unemployed or unskilled person on a fairly low income.

They were also the least likely to use unleaded petrol or energy-saving light bulbs.

YES, SHE'S A GREEN SHOPPER—IT'S THE COLOUR SHE GOES WHEN SHE SEES SOMEONE GETTING A BIGGER BARGAIN

Use two bins . . . a rubbish tip from the Eurocrats

Brussels is talking rubbish – the kind which goes into the kitchen bin.

Eurocrats want Britain's 23 million households to have two dustbins. One would be for organic waste such as vegetables and tea bags and the other for all other household rubbish.

Four million tons of food waste are dumped in landfill sites across Britain every year. The sites leak explosive methane gas and increase global warming.

Brussels is this week discussing proposals for a directive which would force local authorities to increase recycling and the treatment of organic material before disposal.

About 90 per cent of household and industrial rubbish goes into landfill sites, and the clampdown would mean building a network of 70 incinerators across the country to reduce rubbish to ashes before it is buried – at an estimated cost of £5.6 billion.

A tough fight is expected between Brussels and Westminster over who would fund the dustbin revolution. The additional cost of collecting different types of rubbish from separate dustbins would also be considerable.

But environmental groups back the plan, claiming the costs incurred would be outweighed by the jobs created.

A spokesman for Friends of the Earth said: 'If this goes through it could create many jobs for people needed to collect the rubbish and treat it. It is the best way forward – a move away from the throw-away society.'

One Government source said yesterday: 'There would need to be firm guarantees that the cost would not hinder spending plans. The final wording of any legislation will be subject to close scrutiny.'

By David Walmsley

The Department of the Environment said: 'We are unable to comment because we have yet to see the directive.'

Four million tons of food waste are dumped in landfill sites across Britain every year. The sites leak explosive methane gas and increase global warming

British EU Commissioner Sir Leon Brittan is understood to believe that the planned implementation by 2005 is too ambitious a timescale.

EU Environment Commissioner Ritt Bjerregaard is a firm supporter of the plan, over which Britain has no veto.

France, Germany, Austria and Holland already specify incineration of waste before disposal. Scandinavian countries also favour the plan.

However, it could be months before a Commission compromise is ready to be put to Ministers.

Shame of Jack Scout Cove

For the second year running, Jack Scout Cove in Silverdale, Lancashire, has been named the dirtiest beach in Britain.

The Beachwatch '96 operation saw 3,300 volunteers collect 17 tons of rubbish from 203 locations.

Other stretches in the Beachwatch 'dirty dozen' for a second successive year were Port Laing, North Queensferry, Fife (fourth this year), and Outpart East, Harwich, Essex (tenth).

The remainder were: 2 Langness Lighthouse, Isle of Man; 3 East Perwick, Isle of Man; 5 Little Thornwick Bay, East Yorkshire; 6 South Queensferry, West Lothian; 7 Crosby (Burbo Bank), Merseyside; 8 Calf, The Puddle, Isle of Man; 9 Calf, South Harbour, Isle of Man; 11 Strangford Lough (Northern Tip), Co Down; 12 Jennycliff, Devon.

The two cleanest beaches were Mullion, Cornwall, and Pembrey Burrows and Saltings, Carmarthenshire.

*© The Daily Mail
January, 1997*

EU rubbish plan to cost £5.6bn

Toby Helm, EU Correspondent and Charles Clover report

Britain would have to spend £5.6 billion on a network of incinerators to burn most of the nation's rubbish before burying it, under European Commission plans being considered today.

A proposal for a revised EU directive on landfill waste would mean building 70 huge incinerators, at an average cost of £80 million, and would in effect ban all landfill waste – the system used to dispose of 90 per cent of Britain's rubbish – unless it had first been burnt to ash in an incinerator or treated in other ways.

The Government has expressed 'grave concerns' about the directive, which has been drawn up by Ritt Bjerregaard, the Danish environment commissioner.

Officials have made clear that the burning approach cannot be justified on grounds of cost and that no scientific case for it has been made. If the plan is adopted by a majority of the 20 commissioners today it will be considered by EU environment ministers.

Under European Union rules, such a proposal would be subject to a qualified majority vote in the Council of Ministers and Britain would not have a veto.

The worry for the Government is that, because the incineration of waste before disposal in landfill sites is already policy in France, Germany, Austria and Holland, the proposal could be passed.

Mrs Bjerregaard believes that burning waste before dumping it in landfill sites prevents pollution of the land, reduces the amount of space needed for disposal and allows electricity to be generated as a result of the incineration process. It also cuts down on methane, a potent greenhouse gas, which landfills give off.

Britain says that the policy is prohibitively expensive and that research to prove its worth has not been done. Incinerating rubbish would also produce huge quantities of smoke which would be more polluting to the environment than the waste itself.

Under current rules in Britain waste disposal in landfill sites is controlled by lining the sites with sand and clay and mixing the types of waste.

The Government argues that this system works well and says that Mrs Bjerregaard's plan is contrary to an agreement signed by environment ministers in Brussels on Monday, that allowed EU members the 'flexibility to apply the best waste disposal option to meet their particular conditions'.

© The Telegraph plc
London, 1996

European news

Danish battery recovery charges

In Denmark, which incinerates 90 per cent of its residual household waste, the presence of batteries containing heavy metals has been the target of a sustained campaign. Following a covenant between government and industry to recover 75 per cent of nickel-cadmium batteries, a scheme has been introduced which pays collection agencies $US 20,000/tonne of batteries recovered. This is paid for through a $US 1.0 per cell charge.

* * * * * * * *

Survey shows public support for eco-taxes

A recent European Union-wide opinion survey by Eurobarometer shows that almost three out of four people (73 per cent) are in favour of more taxes on products that harm the environment.

* * * * * * * *

Fluorescent lamp recycling

UK company Biffa Waste Services has introduced a nation-wide recycling system for fluorescent lighting tubes. Most of the 80 million lamps sold in Britain each year, containing up to five tonnes of mercury, have ended their days in landfill sites. The new recovery system, named Backtrack by Biffa, is now collecting more than 2,000 lamps each week.

* * * * * * * *

Spain

The Catalan community of Valles Oriental has won almost Pta 1,468 million (approximately £5 million) from the European Union to fund 80 per cent of the costs of building 24 transfer stations for the 43 municipalities. The project will include several treatment plants as well as containers for the separate collection of glass, paper and other materials.

The Balearic island of Menorca has introduced a scheme for separate collection of used cooking oil from hotels and restaurants.

Regional authorities of Castilla La Mancha in central Spain have started work preparing a draft municipal solid waste plan. The first stage is an inventory and diagnosis of the waste, and identification of the main generators.

© Warmer Bulletin, World Resource Foundation
August, 1996

Landfill levy will raise council tax

By Charles Clover, Environment Editor

Householders will have to pay an estimated £5 a year more in council tax as a result of the Landfill Tax to be introduced on 1 October, according to an investigation by *The Daily Telegraph*.

The tax, designed to cut pollution from waste and to promote recycling, is also likely to add costs to all building contracts.

It imposes charges of £7 on every ton of domestic rubbish collected and dumped in landfill sites, a task which costs councils £15–30 per household per year. 'Inert' materials, such as building waste, will attract a levy of £2 a ton.

The extra charge to the householder arises because councils are allowed to pass on the increased cost of disposing of waste through the council tax bill. If they decide not to, or are already rate-capped, they can cut services. Most are expected to pass it on.

The increase in council tax will be highest in areas where councils are worst at recycling, incinerating or composting their waste. Fourteen out of the 15 worst authorities are Labour-controlled, according to the Audit Commission. The joint 15th worst is Sedgefield, in Tony Blair's constituency.

The Landfill Tax will add £154 million to council tax bills according to the Association of Metropolitan Authorities.

This will be offset by a £60 million rebate on employers' national insurance contribution, since a feature of the tax is that it removes tax from employment and places it on pollution. This leaves more than £90 million to be added to council tax.

The result, says the association, will be a charge of £5 for the average band D taxpayer.

The worst bills are likely to be in the next two years, before a Government-led recycling initiative by the packaging industry begins to contribute to councils' recycling costs.

It is clear that some local authorities, such as the London borough of Tower Hamlets which recycles almost no paper, are likely to be wasting hundreds of thousands of pounds a year.

Paper, which makes up around 10 per cent of the average dustbin, has a market value of up to £50 a ton and recycling it is virtually cost-free as it depends only on providing sites for private companies' paper banks. Glass, too, is not being recycled to its fullest extent.

Many councils contacted by *The Telegraph* admitted that they were ill-prepared for the Landfill Tax and had taken little heed of warnings during the last two Budgets that it would increase their costs.

David Staziker, of the Association of County Councils, said: 'We support the principle of reducing waste but we think the Government has got the details wrong. People will either pay more or see something cut, such as expenditure on mending holes in the road.'

The tax is not yet set at a level that would make it cheaper to recycle or incinerate waste, a problem recognised by officials at the Department of the Environment. That level is believed to be around £20 a ton.

Tony Hammond, chairman of the local authority recycling advisory committee, said: 'Everybody is expecting the Chancellor to increase the amount of the landfill Tax in this year's Budget. Realistically, I think we can expect a landfill Tax of £20 a ton within two or three years, whichever party is in power.'

Landfill tax at that level will be needed, according to some experts, for the Government to honour its promise of recycling 25 per cent of the domestic dustbin by 2000.

Knee-deep in rubbish

The Government has set fresh targets for recycling, Ian Murray reports

Britain is faced with too many loads of rubbish. Our sceptered isle is getting clogged up with it. Getting rid of rubbish, of course, is a global problem, but the British are not as good at it as some others. The Germans are undoubtedly the European champions at waste disposal – albeit that they export some of their most noxious stuff for processing.

The difference is that the British have tended to be indifferent to the problem. The average family has tossed a tonne of rubbish into the dustbin each year and relied on the local authority to cart it away. What happened next has not bothered us.

The Germans, having watched their beloved Black Forest dying from the effects of air pollution, have become the green giants of the European Union. The German governments drive environmentally friendly measures through the Council of Ministers, usually dragging a reluctant and protesting Britain in their wake.

It happened over catalytic converters, smoke from power stations and effluent into the North Sea. Now it is happening with rubbish. The EU has set a recycling target of more than 50 per cent of packaging waste by 2001.

In Germany it is perfectly normal for a family to wash up the recyclable rubbish after dinner and put it into the special green dustbin which every household has. In Britain the only major city where the system is under trial is Plymouth, although the small Kentish town of Wye is a national recycling pacesetter.

The figures speak for themselves. In Germany, 30 per cent of household waste is recycled. In Britain, the average is 5 per cent. Only the London Borough of Richmond and Adur District Council in Sussex – both Liberal Democrat controlled – recycle more than 20 per cent. Wye, which salvages 70 per cent of its rubbish, proves we can do it, but this scale of success needs the kind of commitment which is easy only in a small community.

The average family has tossed a tonne of rubbish into the dustbin each year and relied on the local authority to cart it away

---

The Government, having pledged at the Rio Earth summit to recycle 25 per cent of our rubbish by the turn of the century, is now trying to force the pace by wielding its most potent big stick: taxation. From 1 October landfill site operators will have to pay a tax of £2 a tonne for each tonne of dumped inactive waste such as soil or furnace slag, and £7 a tonne for all the unpleasant organic stuff which produces toxic gases and leaches poison into the soil.

The landfill operators will pass this bill on to local authorities, who in turn will pass it on to residents. The Association of County Councils has estimated the total cost to local authorities of the tax will be about £154 million a year, adding about £7 to the average council tax.

The Government undoubtedly hopes that this will concentrate our minds and make us realise that the more we throw away, the more it will cost us.

However, critics of the scheme, including the National Farmers' Union, the Tidy Britain Group, the Country Landowners' Association and the Countryside Movement, fear that the tax will result in a huge increase in fly-dumping, which is illegal.

That is already a danger because we are running out of suitable holes in which to put our rubbish. At the present rate, holes round London will be full within two decades.

Hence the need to make us all recycle more. Just before Christmas, the Government issued its set of targets in a White Paper, *Making Waste Work*. It wants to see the 240 million tonnes of waste presently going to landfill reduced by 60 per cent by 2005. It wants 80 per cent of households to be within easy reach of recycling facilities and 40 per cent of all homes with gardens to have a compost heap by the turn of the century.

According to a series of opinion polls, we are all keen to recycle everything – provided it is made easy for us. Gallup even found recently that four out of ten of us are even prepared to pay a 10 per cent tax on all paper products not made from recycled material.

© The Times
June, 1996

The fungus that ate heavy metal

Waste recycling could become safer thanks to a microscopic mould. David Bradley reports

Landfills are getting fuller and the pressure is on to find alternative methods of recycling our waste. One option is to burn the waste to dust. The dust – or fly ash – can then be compacted down so taking up much less room in a landfill, or better still it can be refabricated into construction materials such as breeze blocks. The trouble is that fly ash from incinerated household and industrial waste is loaded with toxic heavy metals such as cadmium.

Now a group of Swiss scientists think they have come up with a way of extracting the heavy metals from fly ash using a microscopic fungus that grows on trees.

In the October issue of the American journal *Environmental Science and Technology*, Helmut Brandl of the University of Zurich and his team describe how they have used the fungus *Aspergillus niger* – which produces a black mould – to 'mine' the metals from fly ash.

Most simple organisms such as bacteria and single-celled fungi make a collection of chemicals that they secrete into their surroundings to trap the nutrients and minerals – such as iron – which they require to live and grow.

After only one day the fungi they grew had extracted more than half of the metals from the fly ash

---

Once the minerals are trapped the microbes use other molecules on their cell surface to gather up the chemical traps together with their nutrient payload. Brandl and his team realised that if they simply grew the *Aspergillus niger* on fly ash it might produce chemical traps to extract the minerals – the heavy metals – from the waste material.

They found that after only one day the fungi they grew had extracted more than half of the metals from the fly ash – what he calls the 'artificial ore'. They could extract 81 per cent of the cadmium and 32 per cent of the lead in this time. Brandl says that they can make the microbes selective for specific metals by growing strains in higher concentrations of those metals.

The researchers believe this process of 'bio-leaching' allows a safer fly ash to be used for construction materials. It also allows useful metals, such as zinc and aluminium, to be collected for recycling. Brandl has so far carried out back-of-the-envelope calculations on the economics of the process. He reckons it should be less expensive and more energy efficient than the conventional metal evaporation methods sometimes used to extract metals from fly ash.

© *The Guardian*
September, 1996

Recycling in London to get £35m help

The Government will today announce a multi-million-pound grant to boost recycling. The plan is that three years from now the great majority of Londoners will be served by door-to-door collection schemes for waste cans, paper and glass and plastic bottles.

The rapid expansion of recycling should create 1,500 jobs, drastically reduce litter and bring recycling industries into the capital. There are plans for a paper mill using local waste paper, a plastics processing plant and a facility in Hackney which would each year dismantle hundreds of thousands of scrapped computers, which contain small quantities of precious metals.

The strategy comes from a consortium of businesses and almost all of London's 33 boroughs. It has been worked up with support from the Government and its Environment Agency. The boroughs have bid for £35m of City Challenge cash to be spent over the next three years, £10m in the first year. Today, the Secretary of State for the Environment, John Gummer, who has followed the bid closely, will grant a large part of what they want.

Nicky Gavron, a Haringey councillor, said: 'This is about creating jobs in London, bringing in the industries of the future and greening the capital using a partnership between the boroughs, government and business.' She co-chaired the consortium with the former Grand Metropolitan chairman Lord Sheppard of Didgemere.

Recycling has been growing fast in London but it still stands at only 7 per cent of the 3.4 million tons of waste pouring out of the capital each year. Most goes to landfill tips in other counties, but this is becoming more expensive as the dumps fill up, more distant replacements have to

By Nicholas Schoon,
Environment Correspondent

be found and the landfill tax introduced two months ago starts to bite.

Several London boroughs, including Islington, recycle less than 3 per cent

London has Europe's largest recycling resource. But the 7 per cent collected almost all goes outside the capital for processing. For instance, much of the paper goes to a huge new newsprint mill near Maidstone in Kent.

The plan is to raise the amount of materials being collected by recycling from 200,000 tons a year to 500,000 by March in 2000. To achieve that, 30 of the boroughs will have 'multi-material kerbside collection' for household plastic and glass bottles, cans and waste paper.

The consortium also wants to increase composting of garden waste, vegetable peelings and bread six-fold in three years, to 200,000 tons. It foresees 40 per cent of homes with gardens having a compost bin, compared with 10 per cent now.

Robin Murray, a consultant who has done most of the detailed preparation on the scheme, said recycling's greatest pitfall was 'booms and busts', in which those who build up collections suddenly find the market saturated. 'The all-important thing is to build up the supply and demand simultaneously and give both sides of the market long-term confidence,' he said.

Since 1990, the Government's target has been for 25 per cent of household refuse to be recycled by 2000. Bath and Richmond, in south-west London, already achieve that, but it looks unlikely that most councils will hit the target. Several London boroughs, including Islington, recycle less than 3 per cent.

© The Independent
December, 1996

I'VE BEEN TRYING TO RECYCLE HIM FOR YEARS!

Councils recycle much less rubbish

By Charles Clover, Environment Editor

The amount of rubbish recycled by local authorities dropped last year in many areas, including the constituency of John Gummer, the Environment Secretary, according to the Audit Commission.

It said the Government 'will need to address this poor performance' to meet its target of recycling 25 per cent of household rubbish by the year 2000.

The overall percentage of household waste recycled increased in England but fell in Wales. Many councils deteriorated and some recycled nothing at all. Councils in the South and West recycled significantly more than those in the North.

The Commission said the best councils recycled over 20 times more than the worst but it did not appear to cost them any more to do so.

Many councils, which recycled high levels of rubbish, incurred no higher costs per household than councils that recycled much less.

Roger Lilley, for the environmental group Friends of the Earth, said: 'These are shamefully poor figures.'

Councils responsible for significant declines were:
Bolton, Bury, Manchester, Waltham Forest, York, Chester, Rossendale, Wellingborough, Mansfield, Oadby, Daventry, Hereford, Broxbourne, Breckland, Leominster, Hove, Rother, Horsham, West Somerset, Tegnbridge, West Dorset, Exeter, North Cornwall, Winchester and Penwith.

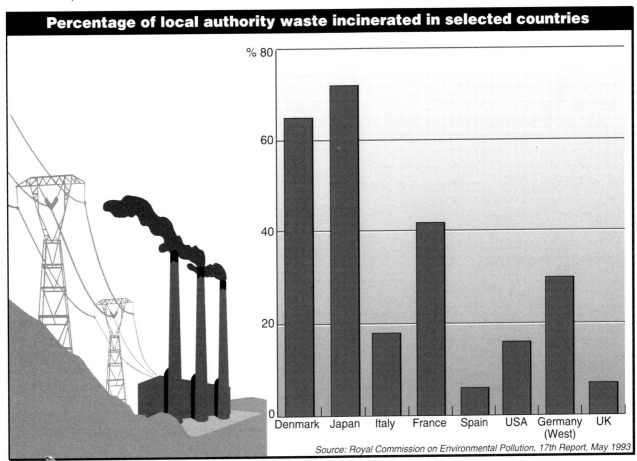

Percentage of local authority waste incinerated in selected countries

Source: Royal Commission on Environmental Pollution, 17th Report, May 1993

The facts

Energy from waste

No matter how green we strive to be, all human activities produce waste. 225 million tonnes of waste were thrown away by Europeans in 1994 – and that's only household waste. It doesn't include all our industrial or commercial waste.

What should we do with our waste?

We can recycle some of it – especially glass, papers such as newsprint, and aluminium for example – but not all waste can be recycled. And recycling is not always economically viable – because recycling some materials can result in a lower-quality product, for which markets are limited, and the costs of collection and recycling can exceed any environmental benefits.

We can bury some of it in landfills – but only some. Landfill cannot go on for ever, and needs careful environmental management. Rising landfill costs mean other disposal methods must be used more.

We can compost some of it – food and garden waste for example. In the Netherlands, Austria and Germany 30% of waste that would otherwise be landfilled is composted.

Or we can burn some of it. This reduces the volume and destroys many of the pollutants in the waste. As waste burns we can harness it for a new use – generating energy.

Energy from waste (EFW) is increasingly seen as an attractive option for waste management. It does three things – treats waste, reduces its volume and generates energy. And it has advantages over other waste management methods and over other energy generation methods.

It is not, however, the only solution, nor is it always the most suitable for a particular town or city. Recovering energy from waste should be seen as one component in an integrated waste management system, to be used alongside recycling, composting, landfill and other measures. In addition, using EFW shouldn't encourage materials producers and users to neglect recycling where it is practical and worthwhile. Nor does it absolve them of their responsibility to reduce waste levels in the first place.

Integrated waste management is the approach advocated by the European Energy from Waste Coalition (EEWC). Within that approach, the EEWC promotes the use of energy from waste – because it is under-utilised, because people don't know enough about it and because it is sometimes misunderstood and wrongly dismissed.

© EEWC

How energy from waste works

How does an EFW plant work?

At a typical EFW plant, waste is unloaded into a large bunker, where it is stored under cover to prevent odour release. It is then fed into a furnace – these generally operate at between 850 and 1350°C. Such high temperatures destroy up to 80% of contaminants such as dioxins and polychlorinated biphenols (PCBS).

Combustion converts the energy content of the waste into heat by creating hot gases. The hot gases are cooled down in a heat exchanger to between 150 and 200°C, and their heat content transferred to water which turns into steam. The steam generates mechanical energy in a steam turbine which is converted into electricity via an alternator.

The energy efficiency of waste compares favourably to other solid fuels. Sorting household waste before it is burned helps increase its energy efficiency. This is because some types of waste have more potential energy in them than others.

Minimally treated waste, where glass, metal and wet organic materials are removed, has an energy potential just under half that of coal. Waste derived from plastics and paper has 80% of coal's energy potential.

Emissions

State-of-the-art control systems keep emissions well within stringent European Union and national regulations, meaning that virtually no odour or harmful substances are released into the atmosphere.

Most of the hot gas that an EFW plant stack emits is simply carbon dioxide and water vapour. Other substances, such as hydrochloric acid, sulphur oxides, nitrous oxides and heavy metals are limited to strictly low levels.

One of the emissions sometimes associated with old-style incineration – dioxins – is far less prevalent at modern EFW plants. In the Netherlands cars and wood fires will produce more dioxins than EFW plants by the year 2000. A modern, average-sized EFW plant emits only 0.25g dioxins a year.

Dioxins are widespread in the environment, so they are also found in waste. EFW is the only waste management technique which actually *reduces* the amount of dioxins in waste. Landfill, recycling and composting have no impact on dioxin levels.

Typically over 80 – 85% of the dioxins in waste are destroyed during combustion. The remainder are stabilised and bound up into the ashes. Less than 1% is emitted back to the atmosphere. In this way modern EFW plants act as an effective 'clean-up' method or 'dioxins sink'.

Ash

After combustion an ash is left behind – approximately 300-350kg from every tonne of waste burned. The vast majority of the ash is grate ash (about 300kg/tonne). This is used extensively as a substrate material for road construction. In Germany and France about 40% of incinerator ash is used for building materials.

A small part of the ash, the residue of the flue gas cleaning process, contains potentially hazardous components. In fact, the better an EFW system is at reducing emissions, the more likely it is that pollutants end up in the ash. However, this ash is stable – the combustion process concentrates potential pollutants from the waste stream, making it simple to deal with. Throughout Europe the ash is carefully disposed of in special landfill sites. A number of pilot projects have looked at physical processes that can extract heavy metals from the ash. Ultimately these metals could be recovered.

© EEWC

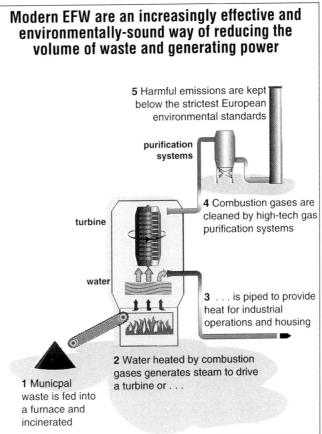

Modern EFW are an increasingly effective and environmentally-sound way of reducing the volume of waste and generating power

5 Harmful emissions are kept below the strictest European environmental standards

purification systems

4 Combustion gases are cleaned by high-tech gas purification systems

turbine

water

3 . . . is piped to provide heat for industrial operations and housing

2 Water heated by combustion gases generates steam to drive a turbine or . . .

1 Municpal waste is fed into a furnace and incinerated

What is energy from waste?

Information from the European Energy from Waste Coalition

What does an EFW plant do?

EFW plants burn municipal solid waste (MSW) and convert the heat energy that is generated into steam and/or electricity. After burning, only around 10% of the original volume of the waste remains as ash.

Using the energy

EFW facilities can use a heat exchanger and boiler to generate steam for community heating networks, particularly for schools, hospitals and publicly owned apartment complexes. For example, the EFW plant at Högdalen in Stockholm burns 220,000 tonnes of waste a year, generating enough steam to fuel 85% of the local district heating system. The waste of five million Parisians provides the Paris District Heating Company with 43% of its needs – enough to heat 198,000 apartments. Producing heat can mean that local residents accept an EFW plant fairly readily because they receive positive benefits from it.

In many EFW plants the exhaust steam is used to drive turbines which generate electricity. An EFW plant in Amsterdam produces 15% of the city's electricity.

Sometimes the heat can be used to dry sludge or distil water for industrial use. The EFW plant in Rotterdam not only produces electricity, but also distils over six million cubic metres of water a year. This is then sold as process water and boiler feed to nearby chemical plants.

Nearly a quarter of EFW plants in Europe produce both heat *and* power. This is known as 'co-generation'. It means the plants can supply local inhabitants with some of the heat and power they need.

Combined heat and power EFW plants are common in Scandinavia and Germany, where it is estimated that the annual waste from seven households can supply the complete annual heat and power needs of one whole household.

One tonne of household waste burnt in a modern EFW plant can:

- generate 300–700kWh of electricity or
- raise the temperature of 30,000 litres of water by 50°C or
- produce 2.5 – 3 tonnes of steam or
- replace 300–350kg of coal or
- provide 15 tonnes of hot water (130°C) for district heating

What is the future for EFW in Europe?

Currently about 33 million tonnes of Europe's municipal solid waste is burned for energy recovery (see figure). In 1994 there were 313 EFW plants in Europe, with an energy capacity equivalent to 11 million tonnes of coal a year.

Both the plant size and the amount of waste treated in this way are expected to rise significantly over the next decade.

Forecasts from the technical consultancy Juniper suggest that waste burned for energy will grow from 33 million tonnes in 1993 to around 83 million tonnes in the year 2000. Conservative estimates suggest that at least 20 new plants will be built every year, other predictions almost double this figure. Currently around 60% of waste-burning facilities recover energy. But this accounts for 90% of the total capacity of all plants. By 2005 energy recovery facilities will be in place in most plants.

Many factors affect the development of EFW in Europe.

- New regulations on incineration emissions, adopted by the European Union in 1989, come into full effect in 1996. These will combine with

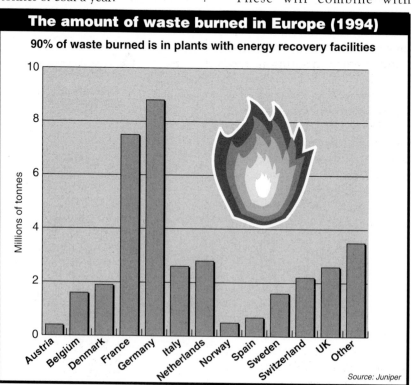

The amount of waste burned in Europe (1994)

90% of waste burned is in plants with energy recovery facilities

Millions of tonnes

Austria, Belgium, Denmark, France, Germany, Italy, Netherlands, Norway, Spain, Sweden, Switzerland, UK, Other

Source: Juniper

stricter national regulations to make EFW more environmentally attractive.

- Older plants, mostly built in the 1970s, are nearing the end of their life, and a natural cycle of replacement and upgrading is under way. The new plants usually have a larger capacity than those they replace.
- European nations are increasingly switching from landfill to EFW as a key component in their waste management strategy. Regulations that prohibit direct landfill of untreated municipal solid waste, along with increased use of landfill taxes, are making landfill less attractive economically.

How environmentally sound is an EFW plant itself?

Many people confuse old-style incinerators with modern state-of-the-art EFW plants. The majority of environmental arguments against old-style incinerators are no longer relevant and cannot be applied to new EFW plants, although the debate over how much modern EFW plants contribute to global warming remains a complex one. Numerous studies have been carried out, most of which conclude that EFW plants have less impact on global warming compared with fossil fuel plants.

Stricter regulations and new technology have made EFW plants amongst the most environmentally sound energy-generating methods available. The maximum level of particulates, for example, has been decreased ten-fold since 1976.

'The substantial environmental benefits of energy from waste plants as part of an integrated waste management strategy – reduced greenhouse gas emissions, energy conservation and diversion of waste from scarce landfill – can be obtained without posing significant health risks.'

Professor Andrew Porteous
Professor of Environmental
Engineering at the
Open University

Detailed facts and figures on emissions and the nature of ash are explained in the EEWC factsheet *How energy from waste works*.

How much does an EFW plant cost?

It should be remembered that cost considerations are never the sole factor in determining an integrated waste management strategy.

In 2005, it is estimated that energy production from EFW plants across Europe will produce the equivalent amount of energy as 27.6 million tonnes of coal a year

---

EFW plants do require more initial investment than conventional fossil fuel power plants. To a certain extent this is due to economies of scale. Fossil fuel plants are on average ten times the size of EFW plants. Furthermore, strict pollution controls require the most up-to-date cleaning and monitoring equipment. About 50–70% of the equipment costs for a new plant goes on technology to comply with environmental regulations, accounting for around 30–50% of the total cost.

Operating costs vary across Europe, from just $70 up to $150 per tonne of waste treated, according to

the Juniper study. There are also considerable differences from one country to another when comparing the operating costs of EFW to other energy-generation methods.

The difference in costs between EFW and alternative forms of power generation varies across Europe, but is most pronounced in countries where EFW is not widely used. Where EFW plants are more common, a number of measures reduce the cost difference. These include:
- levies on fossil fuel energy generation
- subsidies for EFW
- mandated purchase, where electricity companies must buy a proportion of their supply from EFW plants (sometimes at favourably fixed rates)
- long-term contracts
- income from district heating or process steam sold to industry
- pre-sorting the waste to increase its calorific content, thus increasing energy efficiency.

Compared to landfill, EFW is becoming more competitive in some countries. Landfill used to be the cheapest option for managing waste, but now prices are rising due to increasing investment in environmental protection, scarcity of available land, environmental compliance regulations and insurance premiums after the site is closed.

In the case of EFW plants there are various paybacks, both from energy sales and savings from transport, landfill gate fees and a cheaper local source of energy.

Selling electricity can be financially attractive. For example, the Högdalen plant in Stockholm has already recouped 80% of its gross cost through energy sales.

Where government policy means electricity companies must buy a proportion of their supply from EFW plants (sometimes at favourably fixed rates) returns are also good. For example, the Edmonton plant in London made £1.4m from electricity sales in just one month (January 1994). In 2005, it is estimated that energy production from EFW plants across Europe will produce the equivalent amount of energy as 27.6 million tonnes of coal a year.

© EEWC

Waste into energy

Information from the World Resource Foundation

Mixed household waste has around one-third of the energy potential of coal. Recovering energy from the non-recyclable portion of household waste makes economic and environmental sense. The organic content of waste is a renewable source of energy and its recovery displaces fossil fuels, reducing greenhouse gas emissions. The 230 million tonnes of municipal solid waste (MSW) created in Europe each year could meet 5 per cent of Europe's energy needs.

Public acceptance

Despite that, there has often been strong public resistance to energy recovery schemes in many countries. Most of this resistance focuses on waste incineration, rather than on the other, often less well-recognised means of recovering energy from waste.

Plans to build waste treatment and disposal facilities frequently cause dissension where they are proposed. The public may, in principle, accept the need for disposal sites, and generally do support the concept of waste treatment such as composting or recycling, but if plans are announced to build a plant of any kind near their homes, there is frequently opposition.

Because some high-profile environmental groups totally opposed to incineration in any form have become involved in protests about incineration plants rather than other waste facilities, a false impression may be given that only energy recovery plants are opposed.

One reason for public hostility towards waste incineration is the poor performance of old mass burn plants which seldom recovered energy.

Modern waste incinerators are far more likely nowadays to be efficient Energy from Waste (EFW) facilities, where more than half the capital costs pay for energy recovery and pollution control equipment.

It is important to recognise that today's waste plants meet much higher environmental standards than old polluting incinerators, higher too than the coal-fired stations which produce most of the world's electricity.

Dioxins

Dioxin emissions are also cited as a reason for concern about waste incineration.

Dioxins are a family of chemicals, a few of which are thought to be carcinogenic, which are present in the background environment. Dioxins are also formed during manufacturing processes such as that for paper and the metallurgical industries, as well as when organic materials such as coal and wood are burned.

Latest UK predictions suggest that MSW incineration will contribute 0.1 per cent of dioxin emissions to atmosphere by the year 2000.

A risk assessment study has reported that plants operating to current national standards do not pose a significant health risk, irrespective of size and location.

By comparison, vehicle exhausts will contribute 9.5 per cent, domestic coal fires 15.4 per cent and organic materials 35.6 per cent. Recent Europe-wide legislation controlling waste incineration will help to achieve that low level.

Modern EFW units can actually operate as dioxin sinks, trapping and destroying more of the dioxins which were present in the fuel input than they generate as by-products of combustion.

Material versus energy recycling

There is considerable debate as to whether materials recycling and energy recovery should be considered equally valuable, or whether – as many people feel – materials recycling should always be set ahead of energy recovery.

Further controversy concerns the possibility of competition between material and energy recovery options, where combustion is the treatment method. Some opponents are anxious that waste which could be recycled will instead be diverted to EFW schemes because, once a community has built a high-cost plant which needs a minimum throughput, they fear waste minimisation and recycling schemes may be given lower priority.

Methods for recovering energy from waste
- Anaerobic digestion
- Combustion
- Landfill gas recovery
- Pyrolysis and gasification

© *The World Resource Foundation*
October, 1996

Energy from waste

An introduction to its role and environmental benefits.
By Professor Andrew Porteous

Municipal solid waste

Some 18 million tonnes of household waste per year and 12 million tonnes per year of commercial wastes (mainly cardboard, plastic, paper and wood) find their way into the municipal solid waste (MSW) stream.

Britain's 30 million tonnes per year of MSW have to go somewhere. Currently, around 88% of it goes into holes in the ground, known as 'landfill'. Once in the ground these wastes decay, generating gases known as landfill gas, which is principally methane and carbon dioxide. Some of the landfill gas can be burnt as a fuel, but most of it leaks into the atmosphere where it contributes significantly to Britain's greenhouse gas emissions.

It should be noted that landfill is an option of last resort in many European countries. Typically only up to 40% of landfill gas can be captured (based on total gas volume generated from initial waste placement to final stabilisation, perhaps 40–50 years later), leaving 60% to enter the atmosphere as a potent greenhouse gas. Also, polluting leachates (liquids produced from the decomposition of the putrescibles in the waste) can and do enter streams and groundwaters, thereby destroying amenities and/ or valuable water resources for

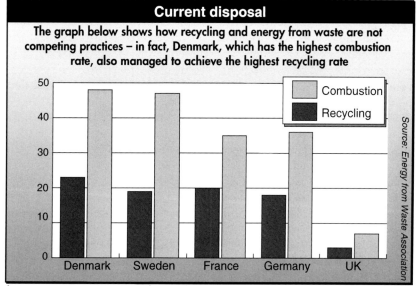

Current disposal

The graph below shows how recycling and energy from waste are not competing practices – in fact, Denmark, which has the highest combustion rate, also managed to achieve the highest recycling rate

future generations unless the landfill is very strictly managed and monitored for up to 50 years after closure.

The remaining 12% of British MSW that is not landfilled is either recycled (roughly 5%) or burnt in specially designed incinerators that may or may not recover the energy released. Most other European countries make much more use of incineration with energy recovery than the UK does, with their plants located in city centres so that the heat produced is used for both power generation and district heating. For example, one-third of central Paris is heated by burning its rubbish, with

energy recovery, in three large incinerators.

The French approach to waste management is termed 'valorisation', i.e. deriving residual value from waste.

The valorisation options (all of equivalent status) are set out below:
- Re-use
- Materials recycling including both mechanical and feedstock recycling
- Energy recovery
- Composting

The main reason for low use of incineration is that the UK has relied on cheap landfill for its waste disposal for far too long, but this is changing.

© Energy from Waste Association

Plastics waste

A source of useful energy

Waste-to-energy conversion processes are increasingly seen as an integral part of good waste management. Waste-to-energy processes will reduce waste volume by typically up to 90% and can be used to generate electricity, steam and/or hot water for industrial or local heating, or to provide energy for large on-site users such as cement producers.

Background

In many countries with the highest environmental standards (e.g. Switzerland and Denmark) where landfill has been restricted, the development and operation of waste-to-energy plants has reached wide commercial application as a major element in their waste management strategy. Operated on a large scale at capacities of over 400,000 tonnes per annum, waste-to-energy plants can generate significant quantities of electrical power, together with lower-value heat for local schemes, whilst operating within tight emission limits.

A modern waste-to-energy plant is capable of generating more than 500 kilowatt-hours of electrical power for every tonne of Municipal Solid Waste feed. If all of Western Europe's Municipal Solid Waste were managed in this manner, 3–4% of domestic electrical power could be generated from our waste.

The average energy value of Municipal Solid Waste is 9 Giga Joules per tonne, compared to about 28 Giga Joules per tonne for industrial coal; put another way, 3 tonnes of Municipal Solid Waste have approximately the same energy value as 1 tonne of coal. By increasing the current Western European level of waste-to-energy conversion from 21% to 50% of the total 120 million tonnes of Municipal Solid Waste, additional energy source materials equivalent to 13 million tonnes of coal could be saved every year. The low-quality heat or steam, also generated, could triple energy savings if fed into the community for district heating.

The contribution made by plastics

The plastics content of Municipal Solid Waste is about 7% by weight. Its high energy value boosts the overall thermal content of the waste stream which helps other, less combustible, elements to burn more readily and efficiently.

Plastics are therefore very important components of Municipal Solid Waste when used for energy recovery. Plastics are mainly derived from oil and consume approximately 4% of the total produced worldwide. The bulk of oil, 86%, is used directly as transportation fuel or to produce energy. Plastics in effect 'borrow' the energy from petroleum or natural gas used to produce them and then 'return' it when used as a fuel in the waste-to-energy conversion.

The UK position

In the UK, approximately 3% of Municipal Solid Waste is used for energy recovery. Successful ventures in this field include the plant in Edmonton, North London, which produces steam to generate electricity. This is then sold to a regional electricity company.

With growing awareness of the potential for developing renewable energy sources as an alternative to fossil fuels, interest is increasing within Government circles and society at large, in using waste for energy generation, opening the way for the UK to join the league of energy-from-waste producers.

If 40% of the UK's 30 million tonnes per annum of Municipal Solid Waste were used as fuel, this could result in the saving of energy equivalent to over 4 million tonnes of coal.

Technology

Incineration, the most widely used waste-to-energy conversion process, is now highly developed and has been significantly improved to reduce emissions. Computer-controlled furnaces, operating at temperatures above 850°C, result in high levels of total combustion efficiency and

conform with the latest EC Directives. This, combined with modern combustion gas cleaning systems, reduces emissions to levels that meet the most stringent environmental regulations.

Through integrated Municipal Solid Waste management schemes, the waste-to-energy process can be further enhanced. Source or 'kerbside' separation of the wet and green fractions, such as garden waste, combined with pre-separation at the incineration plant of inert materials, such as glass and metal (for recovery and recycle), improves the quality of the incineration feed and reduces residues.

The two methods currently used are: mass incineration where collected waste is burned in specially constructed incinerators, and the burning of refuse derived fuel (RDF) in which Municipal Solid Waste has been processed to remove the non-combustibles, such as metal and glass. Coarsely shredded RDF may either be fed directly to conventional power stations, or be further processed and compressed into pellets which can be used in modified boilers as a solid fuel supplement. An added advantage of the RDF route is that it provides an opportunity for the retrieval of materials from Municipal Solid Waste which can be recycled. For example, about 7% of the refuse sent for processing at RDF plants is recovered as ferrous scrap.

The RDF plant at Byker, which is operated by Newcastle City Council, is one of the development plants partially financed by the Department of the Environment and is designed to process 300,000 tonnes per annum of Municipal Solid Waste. Another plant at Pebsham, Hastings, handles 100,000 tonnes of household and commercial waste, and produces 30,000 tonnes of RDF pellets per annum.

The concept of a resource recovery facility (RRF) is another way in which energy may be recovered. This is a centralised waste-processing facility designed to maximise the recovery of marketable secondary products from the waste (glass, ferrous, aluminium, etc.) and/or power. In this way the resource recovery facility underpins material recovery by the production of heat or electricity on site. When the market for secondary products fluctuates, more or less material can be diverted to combustion in order to maintain economic operation.

Waste processing and combustion could also be fully integrated with other treatment facilities. For example, the anaerobic digestion of kitchen waste and other organic rich 'fines' contained in Municipal Solid Waste can generate methane gas for further energy recovery. In this way, the concept of centralised resource recovery could be adapted to local requirements and opportunities.

Emission controls

In modern plants, emissions from the burning of solid waste are controlled by maintaining proper combustion conditions, and by the use of comprehensive pollution control devices, such as electrostatic precipitators, bag filters and acid gas scrubbers.

In modern incinerators, furnaces are designed to achieve a minimum temperature of 850°C with a residence time of at least 2 seconds. Studies show that these conditions are necessary, not only to promote efficient conversion of almost all carbon content to carbon dioxide, but also to minimise the emission of dioxins.

Heavy metals, in general, are concentrated in the ash and disposed of either as hazardous waste, or in appropriate installations where they can be retrieved and recovered.

© British Plastics Federation
Autumn, 1992

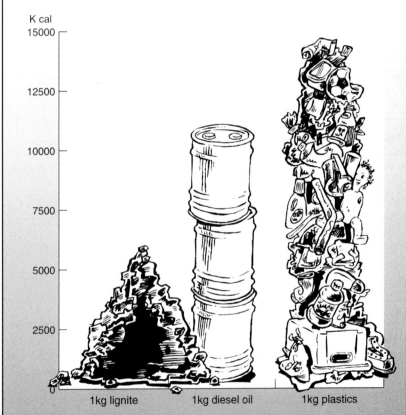

Energy from waste

Reduction, re-use and recycling are not the only waste management options. Waste can also be burnt and the heat energy produced used as heating or in power generation. This diagrm shows the heat energy content equivalent of 1kg of lignite, diesel oil and plastics.

K cal

1kg lignite   1kg diesel oil   1kg plastics

The energy content of waste plastics can be equivalent to coal or oil. And it is the plastics in waste, along with other combustible items, which makes the waste such a good fuel. The 8% plastics content produces 30% of the heat energy released. Already across Europe over 27 million tonnes of waste are burnt each year to produce useful heat energy.

Source: Association of Plastics Manufacturers in Europe

INDEX

ADDITIONAL RESOURCES

Association of Plastics Manufacturers in Europe
Box 3 Avenue E Van
Nieuwenhyse 4
B-1160 BRUSSELS
Tel: 00 32 2 672 8259
Fax: 00 32 2 675 3935

British Plastics Federation (BPF)
6 Bath Place
Rivington Street
London, EC2A 3JE
Tel: 0171 457 5000
The BPF is the premier trade federation for the UK plastics industry. A wide variety of sectors are represented including suppliers of polymers, additives, machinery, processors and equipment.

EarthKind
Bounds Green Road
London N22 4EU
Tel: 0181 889 1595
Fax: 0181 881 7662
Works to bring a new perspective to the green scene by uniting environmental and animal welfare issues. Produces the publication *Living World*.

Energy from Waste Association
26 Spring Street
London
W2 1JA
Tel: 0171 402 7110
Fax: 0171 402 7115
Represents operators and developers of energy from waste plants. Promotes the benefits of recovering energy from waste. Produces booklet, reports and other information on issues relating to energy from waste.

European Energy From Waste Coalition (EEWC)
rue d'Arlon 50
1040 Brussels
Belgium
Tel: 00 32 2 280 1790
Fax: 00 32 2 280 1883
Produces booklet, reports and other information on issues relating to energy from waste.

Friends of the Earth
26-28 Underwood Street
London N1 7JQ
Tel: 0171 490 1555
Fax: 0171 490 0881
As an independent environmental group, Friends of the Earth publishes a comprehensive range of leaflets, books and in-depth briefings and reports. Ask for their publications list publications.

Marine Conservation Society
9 Gloucester Road
Ross on Wye
Herefordshire HR9 5BU
Tel: 01989 566 017
Information on conservation of water habitats. Please send an SAE with any requests for information. The *Beachwatch '96* report is £5 and their factsheets are 50 pence each.

National Recycling Forum
Gresham House
24 Holborn Viaduct
London EC1A 2BN
Tel: 0171 248 1412
Fax: 0171 248 1404
Ask for their publications list publications.

Pulp and Paper Information Centre
Papermakers House
Rivenhall Road
Westleigh
Swindon SN5 7BD
Tel: 01793 886086
Produces a wide range of free factsheets and other information. Ask for their publications list publications.

The Wildlife Trusts
The Green
Witham Park
Waterside South
Lincoln LN5 7JR
Tel: 01522 544 400
Produces a wide range of free factsheets and other information. Ask for their publications list publications.

Tidy Britain Group
The Pier
Wigan
Lincs WN3 4EX
Tel: 01942 824620
Fax: 01942 824778
Protects and enhances the amenities of town and country particularly by controling the prevention and control of litter. Produces publications.

Waste Watch
Gresham House
24 Holborn Viaduct
London EC1A 2BN
Tel: 0171 248 1818
Fax: 0171 245 9719
Promotes and supports action on waste reduction, re-use and recycling in the community. Produces a wide range of free factsheets and other information. Ask for their publications list publications.

Women's Environmental Network (WEN)
87 Worship Street
London EC2A 4BE
Tel: 0171 247 3327
Fax: 0171 247 4740
The Women's Environmental Network is one of Britain's leading environmental charities. Their aim is to educate, inform and empower women who care about the environment. The WEN Information Department answers inquiries, produces newsletters, briefings, papers and other information related to women and the environment. Please send a large SAE for general WEN information or information on waste.

World Resource Foundation
Bridge House
High Street, Tonbridge
Kent TN9 1DP
Tel: 01732 368 333
Fax: 01732 368 337
Produce the magazine *Warmer Bulletin*.

ACKNOWLEDGEMENTS

The publisher is grateful for permission to reproduce the following material.

While every care has been taken to trace and acknowledge copyright, the publisher tenders its apology for any accidental infringement or where copyright has proved untraceable. The publisher would be pleased to come to a suitable arrangement in any such case with the rightful owner.

Chapter One: Recycling

What did you throw out this week?, © HMSO, Reproduced by kind permission of Her Majesty's Stationery Office, February 1996, *Research and surveys*, © Tidy Britain Group, *Typical household waste analysis*, © Warren Spring Laboratory, *Don't throw it all away*, © Friends of the Earth, *The litter collection*, © Tidy Britain Group, August 1996, *Why recycle?*, © Waste Watch, May 1995, *Another world first for recycling*, © National Recycling Forum, *Making waste work*, © HMSO, Reproduced by kind permission of Her Majesty's Stationery Office, December 1995, *The WEN guide to waste and how to prevent it*, © The Women's Environmental Network, February 1996, *One week's waste from a family of two adults and two pre-teen children*, © Waste Watch, *The law against littering*, © Tidy Britain Group, *Beachwatch 1996*, © Marine Conservation Society Ltd, 1996, *Tourists blamed for bulk of beach debris*, © The Independent, January 1997, *Breakdown of plastic debris items*, © Marine Conservation Society, *Looking at litter and recycling*, © Wildlife Trusts, *New UK Government strategy on recycling*, © HMSO, Reproduced by kind permission of Her Majesty's Stationery Office, February 1997, *The marine debris scandal*, © The Living World, Winter 1996, *New agreement achieved on waste dumping at sea*, © MAFF, November 1996,

Healthy seas are vital, © Marine Conservation Society, *Recycling – the facts*, © Pulp & Paper Information Centre, *EC Directive on Packaging and Packaging Waste*, © Waste Watch, *The green shopper 'is alive and well'*, © The Telegraph Plc, London 1996, *Use two bins . . . a rubbish tip from the Eurocrats*, © The Daily Mail, January 1997, *EU rubbish plan to cost £5.6 bn*, © The Telegraph Plc, London 1996, *European news*, © World Resource Foundation, August 1996, *Landfill levy will raise council tax*, © The Telegraph Plc, London 1996, *Knee-deep in rubbish*, © Times Newspapers Ltd, June 1996, *The fungus that ate heavy metal*, © The Guardian, September 1996, *Recycling in London to get £35m help*, © The Independent, December 1996, *Councils recycle much less rubbish*, © The Telegraph Plc, London 1996.

Chapter Two: Energy from waste

The facts, © EEWC, *How energy from waste works*, © EEWC, *What is energy from waste?*, © EEWC, *Waste into energy*, © The World Resource Foundation, October 1996, *Energy from waste*, © Energy from Waste Association, *Plastics waste*, © British Plastics Federation.

Photographs and Illustrations

Pages 1, 8, 22, 27, 29, 30: Ken Pyne, pages 4, 6, 25: Andrew Smith / Folio Collective, pages 16, 23, 36: Katherine Fleming / Folio Collective.

Craig Donnellan
Cambridge
January, 1997